ELECTRICAL MOTOR CONTROLS
for Integrated Systems

Applications Manual

AMERICAN TECHNICAL PUBLISHERS, INC.
HOMEWOOD, ILLINOIS 60430-4600

Glen A. Mazur

Electrical Motor Controls for Integrated Systems Applications Manual contains procedures commonly practiced in industry and the trade. Specific procedures vary with each task and must be performed by a qualified person. For maximum safety, always refer to specific manufacturer recommendations, insurance regulations, specific job site and plant procedures, applicable federal, state, and local regulations, and any authority having jurisdiction. The material contained is intended to be an educational resource for the user. American Technical Publishers, Inc., assumes no responsibility or liability in connection with this material or its use by any individual or organization.

© 2005 by American Technical Publishers, Inc.
All rights reserved

1 2 3 4 5 6 7 8 9 – 05 – 9 8 7 6 5 4 3 2

Printed in the United States of America

ISBN 978-0-8269-1214-5

 This book is printed on 30% recycled paper.

electrical motor controls for Integrated Systems

Contents

1 ELECTRICAL QUANTITIES AND CIRCUITS

Application 1-1	Electrical Prefixes	1
Application 1-2	Using Ohm's Law and the Power Formula	2
Application 1-3	Resistance in Series Circuits	2
Application 1-4	Resistance in Parallel Circuits	3
Application 1-5	Resistance in Series/Parallel Circuits	4
Application 1-6	Resistor Color Coding	5
Application 1-7	Finding Total Resistance of Resistors Connected in Parallel Using a Calculator	6
Activity 1-1		7
Activity 1-2		7
Activity 1-3		9
Activity 1-4		10
Activity 1-5		11
Activity 1-6		12
Activity 1-7		13

2 ELECTRICAL TOOLS AND TEST INSTRUMENTS

Application 2-1	Multimeter Use	15
Application 2-2	Temperature Conversion	18
Application 2-3	Reading Graphs	19
Application 2-4	Applying Electrical Principles when Troubleshooting	20
Activity 2-1		21
Activity 2-2		22
Activity 2-3		22
Activity 2-4		24

3 ELECTRICAL SAFETY

Application 3-1	Electrical Glove Selection	25
Application 3-2	Grounding	26
Application 3-3	Safety Label Information	26
Application 3-4	Troubleshooting Ground Problems	27
Activity 3-1		29
Activity 3-2		30
Activity 3-3		31
Activity 3-4		32

4 ELECTRICAL SYMBOLS AND DIAGRAMS

Application 4-1	Symbol and Abbreviation Identification	33
Application 4-2	Wiring and Line Diagrams	34
Application 4-3	Allowable Voltage Drop across Conductors	35
Application 4-4	Troubleshooting Power and Control Circuits	38
Activity 4-1		39
Activity 4-2		43
Activity 4-3		44
Activity 4-4		45

5 LOGIC APPLIED TO LINE DIAGRAMS

Application 5-1	Numerical Cross-Reference Numbers	47
Application 5-2	Assigning Wire-Reference (Terminal) Numbers	48
Application 5-3	Manufacturer Terminal Numbers	49
Application 5-4	Wiring Control Panels	50
Application 5-5	Basic Switching Logic	50
Application 5-6	Troubleshooting Control Circuits	51
Activity 5-1		55
Activity 5-2		56
Activity 5-3		60
Activity 5-4		61
Activity 5-5		62
Activity 5-6		64

6 SOLENOIDS, DC GENERATORS, AND DC MOTORS

Application 6-1	Solenoids	69
Application 6-2	Directional Control Valves	71
Application 6-3	Fluid Power Color Coding	73
Application 6-4	Troubleshooting DC Motor Circuits	74

Activity 6-1	75
Activity 6-2	76
Activity 6-3	77
Activity 6-4	80

7 AC GENERATORS, TRANSFORMERS, AND AC MOTORS

Application 7-1	AC Generators	81
Application 7-2	Transformers	83
Application 7-3	Troubleshooting AC Motors and Motor Circuits	84

Activity 7-1	85
Activity 7-2	86
Activity 7-3	88

8 CONTACTORS AND MOTOR STARTERS

Application 8-1	Ambient Temperature Compensation with Overloads	91
Application 8-2	Overload Trip Time	92
Application 8-3	Overload Heater Size	93
Application 8-4	Contactor and Motor Starter Ratings	93
Application 8-5	Sizing Motor Protection, Motor Starters, and Wire	94
Application 8-6	Motor Starter Replacement Parts	95
Application 8-7	Motor Drives	96
Application 8-8	Troubleshooting Contactors	96

Activity 8-1	97
Activity 8-2	97
Activity 8-3	98
Activity 8-4	99
Activity 8-5	99
Activity 8-6	101
Activity 8-7	102
Activity 8-8	104

9 CONTROL DEVICES

Application 9-1	Enclosure Selection	107
Application 9-2	Alternating Motor Control	109
Application 9-3	Level Control	110
Application 9-4	Temperature Control	111
Application 9-5	Selecting Blowers and Exhaust Fans	111
Application 9-6	Troubleshooting Control Device Circuits	113

Activity 9-1	115
Activity 9-2	115
Activity 9-3	117
Activity 9-4	118
Activity 9-5	119
Activity 9-6	120

10 REVERSING MOTOR CIRCUITS

Application 10-1	Reversing Motor Circuits	123
Application 10-2	Reversing Three-Phase Motors	124
Application 10-3	Reversing Single-Phase Motors	124
Application 10-4	Reversing Dual-Voltage Motors	126
Application 10-5	Reversing DC Motors	126
Application 10-6	Troubleshooting Reversing Motor Circuits	127
Application 10-7	Hard Wiring Reversing Circuits	128
Application 10-8	Reversing Circuits and Terminal Strips	129
Application 10-9	Reversing Circuits and PLCs	131

Activity 10-1	133
Activity 10-2	134
Activity 10-3	135
Activity 10-4	136
Activity 10-5	140
Activity 10-6	141
Activity 10-7	142
Activity 10-8	144
Activity 10-9	147

11 POWER DISTRIBUTION SYSTEMS

Application 11-1	Wye and Delta Transformer Configurations	149
Application 11-2	Motor Control Centers	149
Application 11-3	Busway Systems	152
Application 11-4	Troubleshooting Power Circuits	153
Application 11-5	120/240 V, Single-Phase Systems	154
Application 11-6	120/208 V, Three-Phase Systems	155
Application 11-7	120/240 V, Three-Phase Systems	156
Application 11-8	277/480 V, Three-Phase Systems	157
Application 11-9	Plug and Receptacle Configurations and Ratings	158

Activity 11-1	159
Activity 11-2	163
Activity 11-3	163
Activity 11-4	166
Activity 11-5	168
Activity 11-6	169
Activity 11-7	170
Activity 11-8	171
Activity 11-9	172

12 SOLID-STATE DEVICES AND SYSTEM INTEGRATION

Application 12-1	Electronic Device Symbols	173
Application 12-2	Digital Circuit Logic Functions	173
Application 12-3	Combination Logic Circuits	175
Application 12-4	Solid-State Relays and Switches	177
Application 12-5	Photovoltaic Cells	179
Application 12-6	Troubleshooting Digital Circuits	179

Activity 12-1	181
Activity 12-2	182
Activity 12-3	184
Activity 12-4	186
Activity 12-5	188
Activity 12-6	189

13 TIMERS AND COUNTERS

Application 13-1	ON-Delay Timer Applications	191
Application 13-2	OFF-Delay Timer Applications	192
Application 13-3	One-Shot Timer Applications	193
Application 13-4	Recycle Timer Applications	194
Application 13-5	Combination Timing Logic Applications	195
Application 13-6	Selecting and Setting Timers	196
Application 13-7	Troubleshooting Timer Circuits	196

Activity 13-1	197
Activity 13-2	198
Activity 13-3	199
Activity 13-4	200
Activity 13-5	201
Activity 13-6	202
Activity 13-7	205

14 RELAYS AND SOLID-STATE STARTERS

Application 14-1:	Relays	207
Application 14-2:	Heat Sink Selection	209
Application 14-3:	Heat Sink Installation	211
Application 14-4:	Solid-State Relay Installation	211
Application 14-5:	Troubleshooting at Motor Starters	212

Activity 14-1	213
Activity 14-2	216
Activity 14-3	216
Activity 14-4	217
Activity 14-5	218

15 SENSING DEVICES AND CONTROLS

Application 15-1	Proximity Sensors	219
Application 15-2	Proximity Sensor Installation	221
Application 15-3	Determining Activating Frequency	222
Application 15-4	Applying Photoelectric Sensors	224
Application 15-5	Troubleshooting Photoelectric Sensors	224

Activity 15-1	225
Activity 15-2	226
Activity 15-3	227
Activity 15-4	228
Activity 15-5	229

16 PROGRAMMABLE CONTROLLERS

Application 16-1	Programmable Controller Input and Output Identification ___ 231	Activity 16-1 ___ 237
Application 16-2	Programmable Controller Input and Output Connections ___ 232	Activity 16-2 ___ 238
Application 16-3	Alarm Output Connection ___ 233	Activity 16-3 ___ 239
Application 16-4	Developing PLC Programmable Circuits ___ 234	Activity 16-4 ___ 240
Application 16-5	Troubleshooting PLC Inputs and Outputs ___ 235	Activity 16-5 ___ 244

17 REDUCED-VOLTAGE STARTING

Application 17-1	Primary Resistor Reduced-Voltage Starting ___ 245	Activity 17-1 ___ 249
Application 17-2	Part-Winding Reduced-Voltage Starting ___ 245	Activity 17-2 ___ 250
Application 17-3	Autotransformer Reduced-Voltage Starting ___ 246	Activity 17-3 ___ 251
Application 17-4	Wye/Delta Reduced-Voltage Starting ___ 247	Activity 17-4 ___ 252
Application 17-5	Closed Transition Reduced-Voltage Starting ___ 248	Activity 17-5 ___ 253
Application 17-6	Troubleshooting Reduced-Voltage Starting Circuits ___ 248	Activity 17-6 ___ 254

18 ACCELERATING AND DECELERATING METHODS

Application 18-1	One-Direction Motor Plugging ___ 257	Activity 18-1 ___ 263
Application 18-2	Two-Direction Motor Plugging ___ 258	Activity 18-2 ___ 264
Application 18-3	Two-Speed Separate Winding Motors ___ 259	Activity 18-3 ___ 264
Application 18-4	Two-Speed Consequent Pole Motors ___ 260	Activity 18-4 ___ 265
Application 18-5	Motor Torque and Horsepower ___ 261	Activity 18-5 ___ 266
Application 18-6	Troubleshooting Two-Speed Circuits ___ 261	Activity 18-6 ___ 268
Application 18-7	Troubleshooting Two-Direction Plugging Circuits ___ 262	Activity 18-7 ___ 269
Application 18-8	Troubleshooting Electronic Braking Circuits ___ 262	Activity 18-8 ___ 270

19 PREVENTIVE MAINTENANCE AND TROUBLESHOOTING

Application 19-1	Conveyor Drive Methods ___ 271	Activity 19-1 ___ 279
Application 19-2	Motor Coupling Selection ___ 273	Activity 19-2 ___ 280
Application 19-3	Extension Cord Selection ___ 275	Activity 19-3 ___ 280
Application 19-4	Load Variations ___ 277	Activity 19-4 ___ 282

A APPENDIX ___ 285

Introduction

Electrical Motor Controls for Integrated Systems Applications Manual includes hands-on application information that expands on content presented in *Electrical Motor Controls for Integrated Systems*, 3rd Edition. The applications manual consists of applications and activities. The applications present technical, manufacturing, and troubleshooting data as it appears in service manuals used by industrial electricians. Selected motor control topics highlight the proper use, sizing, connection, and troubleshooting of electrical motor control devices.

The activities reinforce the understanding of the topics presented and help prepare students for proper ordering, installation, maintenance, and troubleshooting of electrical motor control devices and circuits. Questions in the activities include identification, completion, calculations, short answer, and illustrated answers. All answers to the questions are given in the *Electrical Motor Controls for Integrated Systems Applications Manual Answer Key*.

The *Electrical Motor Controls for Integrated Systems Applications Manual* may be used independently or in conjunction with *Electrical Motor Controls for Integrated Systems*, 3rd Edition, for a more in-depth study. A comprehensive Appendix provides useful information in an easy-to-find format. Four certificates of completion are included in the back of the book. The certificates can be used to document instruction provided and activities completed in the specific areas indicated on each certificate.

The Publisher

Electrical Quantities and Circuits

Applications 1

Application 1-1: Electrical Prefixes

Prefixes

Measured or calculated electrical units may be large or small. For example, solid-state devices may have a current draw of less than 0.000001 A (amperes). In an industrial plant that melts aluminum, power greater than 100,000 W (watts) may be used.

To avoid long expressions, prefixes are used to indicate units that are smaller or larger than the base unit. For example, 0.000001 A equals 1 µA (microampere), and 100,000 W is equal to 100 kW (kilowatts). **See Common Prefixes.**

COMMON PREFIXES		
Symbol	Prefix	Equivalent
G	giga	1,000,000,000
M	mega	1,000,000
k	kilo	1000
base unit	—	1
m	milli	0.001
µ	micro	0.000001
n	nano	0.000000001
p	pico	0.000000000001

Converting Units

To convert between different units, move the decimal point to the left or right, depending on the unit. **See Conversion Table.**

> *Example: Converting Units*
> Convert 0.000001 A to simplest terms.
> Move the decimal point six places to the right to obtain 1.0 µA (from Conversion Table).
> 0.000001 A = 1.0 µA or 1 µA

CONVERSION TABLE						
Initial Units	Final Units					
	giga	mega	kilo	base unit	milli	micro	nano
giga		3R	6R	9R	12R	15R	18R
mega	3L		3R	6R	9R	12R	15R
kilo	6L	3L		3R	6R	9R	12R
base unit	9L	6L	3L		3R	6R	9R
milli	12L	9L	6L	3L		3R	6R
micro	15L	12L	9L	6L	3L		3R
nano	18L	15L	12L	9L	6L	3L	

R = move the decimal point to the right.
L = move the decimal point to the left.

Common Electrical Quantities

Abbreviations are used for common electrical quantities to simplify their expression. **See Common Electrical Quantities.**

> *Example: Electrical Abbreviations*
> Abbreviate the following electrical terms.
> 120 milliwatts = 120 mW
> 120 watts = 120 W
> 50 farads = 50 F
> 120 kilovolts = 120 kV
> 10 amperes = 10 A

COMMON ELECTRICAL QUANTITIES		
Variable	Name	Unit of Measure and Abbreviation
E	voltage	volts — E
I	current	amperes — A
R	resistance	ohms — Ω
P	power	watts — W
P	power (apparent)	volt-amps — VA
C	capacitance	farads — F
L	inductance	henrys — H
Z	impedance	ohms — Ω

1

Application 1-2: Using Ohm's Law and the Power Formula

Ohm's Law

Ohm's law is the relationship between the voltage, current, and resistance in an electrical circuit. Ohm's law states that current in a circuit is proportional to the voltage and inversely proportional to the resistance. Ohm's law is written $I = \dfrac{E}{R}$, $R = \dfrac{E}{I}$, and $E = R \times I$.

Power Formula

The *power formula* is the relationship between the voltage, current, and power in an electrical circuit. The power formula states that the power in a circuit is equal to the voltage times the current. The power formula is written $P = E \times I$, $E = \dfrac{P}{I}$, and $I = \dfrac{P}{E}$.

Any value in these relationships can be found using Ohm's law and the power formula. **See Ohm's Law and Power Formula.**

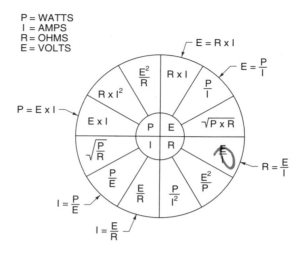

OHM'S LAW AND POWER FORMULA

Example: Finding Current Using Ohm's Law
A circuit has a voltage supply of 120 V and a resistance of 60 Ω. Find the current in the circuit.
$I = \dfrac{E}{R}$
$I = \dfrac{120}{60}$
$I = \mathbf{2\ A}$

Example: Finding Power Using Power Formula
A load that draws 5 A is connected to a 240 V power supply. Find the power in the circuit.
$P = E \times I$
$P = 240 \times 5$
$P = \mathbf{1200\ W}$

Application 1-3: Resistance in Series Circuits

Series Circuits

A *series circuit* is a circuit that contains two or more loads and one path through which current flows. The total resistance of the loads (resistors) connected in series is equal to the sum of the individual resistances. **See Series Circuit.**

SERIES CIRCUIT

The total resistance of a series circuit is found by applying the formula:
$R_T = R_1 + R_2 + R_3 + \ldots$
where
R_T = total resistance (in Ω)
R_1 = resistance 1 (in Ω)
R_2 = resistance 2 (in Ω)
R_3 = resistance 3 (in Ω)

Example: Finding Total Resistance—Series Circuit
A circuit has four resistors of 10 Ω, 55 Ω, 100 Ω, and 800 Ω connected in series. Find the total resistance in the circuit.
$R_T = R_1 + R_2 + R_3 + R_4$
$R_T = 10 + 55 + 100 + 800$
$R_T = \mathbf{965\ \Omega}$

Application 1-4: Resistance in Parallel Circuits

Parallel Circuits

A *parallel circuit* is a circuit that contains two or more loads and has more than one path through which current flows. **See Parallel Circuit.** The total resistance of a parallel circuit with two resistors is found by applying the following formula:

$$R_T = \frac{R_1 \times R_2}{R_1 + R_2}$$

The total resistance of a parallel circuit with three or more resistors is found by applying the following formula:

$$R_T = \frac{1}{\dfrac{1}{R_1} + \dfrac{1}{R_2} + \dfrac{1}{R_3} + \ldots}$$

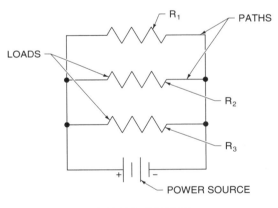

PARALLEL CIRCUIT

Example: Finding Total Resistance—Parallel Circuit
A circuit has two resistors of 50 Ω and 200 Ω connected in parallel. Find the total resistance in the circuit.

$$R_T = \frac{R_1 \times R_2}{R_1 + R_2}$$

$$R_T = \frac{50 \times 200}{50 + 200}$$

$$R_T = \frac{10{,}000}{250}$$

$$R_T = 40 \text{ Ω}$$

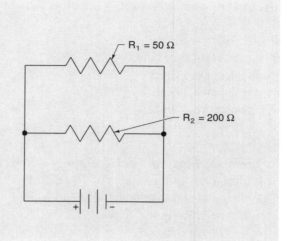

Application 1-5: Resistance in Series/Parallel Circuits

Series/Parallel Circuits

Most electrical circuits are combinations of series and parallel circuits. The total resistance of a series/parallel circuit is found by calculating the equivalent resistances of the parallel circuit(s) and adding the value to the resistance of the loads connected in series. **See Series/Parallel Circuit.** The total resistance in a series/parallel circuit is found by applying the following formula:

$$R_T = \frac{R_{P1} \times R_{P2}}{R_{P1} + R_{P2}} + R_{S1} + R_{S2} + \ldots$$

where
R_T = total resistance (in Ω)
R_{P1} = parallel resistance 1 (in Ω)
R_{P2} = parallel resistance 2 (in Ω)
R_{S1} = series resistance 1 (in Ω)
R_{S2} = series resistance 2 (in Ω)

SERIES/PARALLEL CIRCUIT

Example: Finding Total Resistance—Series/Parallel Circuit
A circuit has two resistors of 150 Ω and 300 Ω connected in parallel and three resistors of 75 Ω, 50 Ω, and 25 Ω connected in series. Find the total resistance in the circuit.

$$R_T = \frac{R_{P1} \times R_{P2}}{R_{P1} + R_{P2}} + R_{S1} + R_{S2} + R_{S3}$$

$$R_T = \frac{150 \times 300}{150 + 300} + 75 + 50 + 25$$

$$R_T = \frac{45{,}000}{450} + 150$$

$$R_T = 100 + 150$$

$$R_T = \mathbf{250\ \Omega}$$

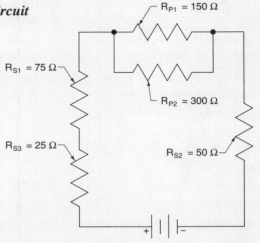

Application 1-6: Resistor Color Coding

Resistors

A *resistor* is a device that limits the current flowing in an electrical circuit. Small resistors use color bands to represent their resistance value. The first two color bands represent the first two digits in the value of the resistor. The third color band (multiplier) indicates the number of zeros that must be added to the first two digits. The fourth band (tolerance) indicates how far the actual measured value can be from the coded value. **See Resistor Color Codes.**

RESISTOR COLOR CODES			
Color	1st Number	2nd Number	Multiplier	Tolerance (%)
Black	0	0	1	0
Brown	1	1	10	—
Red	2	2	100	—
Orange	3	3	1000	—
Yellow	4	4	10,000	—
Green	5	5	100,000	—
Blue	6	6	1,000,000	—
Violet	7	7	10,000,000	—
Gray	8	8	100,000,000	—
White	9	9	1,000,000,000	—
Gold	—	—	0.1	5
Silver	—	—	0.01	10
None	—	—	0	20

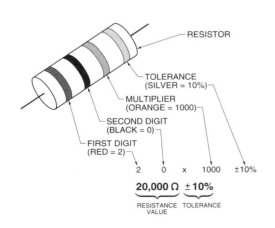

Example: Finding Resistor Value
A resistor with red, black, orange, and silver color bands has a resistance value of 20,000 Ω (20 kΩ) ±10%.

Application 1-7: Finding Total Resistance of Resistors Connected in Parallel Using a Calculator

Finding Total Resistance Using a Calculator

The total resistance of a parallel circuit containing three or more resistors is found by applying the following formula:

$$R_T = \frac{1}{\frac{1}{R_1} + \frac{1}{R_2} + \frac{1}{R_3} + \ldots}$$

A calculator can be used to easily apply this formula when determining the total resistance of a parallel circuit containing three or more resistors. **See Calculator.** The total resistance in a parallel circuit containing three resistors is found using a calculator by applying the following procedure:

1. Clear the calculator so it reads 0. Ensure it does not read 0^M.
2. Enter the keystrokes 1, ÷, R1 value, and M+ to enter the first resistor value into memory.
3. Enter the keystrokes 1, ÷, R2 value, and M+ to add the second resistor value into memory.
4. Enter the keystrokes 1, ÷, R3 value, and M+ to add the third resistor value into memory.
5. Enter the keystrokes 1, ÷, MR (or RM), and =.

The calculator displays the total resistance of the three resistors connected in parallel.

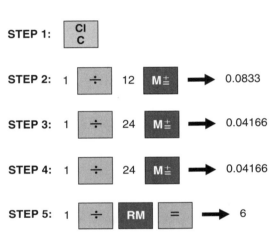

CALCULATOR

Example: Finding Total Resistance of Resistors Connected in Parallel Using a Calculator
A circuit has three resistors of 12 Ω, 24 Ω, and 24 Ω connected in parallel. Find the total resistance of the circuit using a calculator.

1. Clear the calculator so it reads 0. Ensure it does not read 0^M.
2. Enter the keystrokes 1, ÷, 12, and M+ to enter the 12 Ω resistor value into memory. The calculator should read 0.08333.
3. Enter the keystrokes 1, ÷, 24, and M+ to enter the first 24 Ω resistor value into memory. The calculator should read 0.04166.
4. Enter the keystrokes 1, ÷, 24, and M+ to enter the second 24 Ω resistor value into memory. The calculator should read 0.04166.
5. Enter the keystrokes 1, ÷, MR (or RM), and =. The calculator should read 6.

The total resistance of the parallel circuit is **6 Ω**.

Electrical Quantities and Circuits

Activities 1

Name _____ Date _____

Activity 1-1: Electrical Prefixes

State each value in its simplest form.

_____ 1. 0.045 V = ___   _____ 11. 0.00360 A = ___

_____ 2. 22,000 Ω = ___   _____ 12. 3,300,000 Ω = ___

_____ 3. 0.006 A = ___   _____ 13. 0.00005 W = ___

_____ 4. 0.0004 V = ___   _____ 14. 0.600000 V = ___

_____ 5. 0.00000052 A = ___   _____ 15. 0.000000002 W = ___

_____ 6. 21,000,000,000 W = ___ _____ 16. 35,100,000 W = ___

_____ 7. 0.00000000004 V = ___ _____ 17. 3005 V = ___

_____ 8. 0.005 V = ___   _____ 18. 1000.5 A = ___

_____ 9. 2000 V = ___   _____ 19. 10.050 A = ___

_____ 10. 0.00000003 A = ___   _____ 20. 0.1001 W = ___

Activity 1-2: Using Ohm's Law and the Power Formula

Solve the problems using Ohm's law or the power formula.

___2 × 120 =_____ 1. $E = \underline{240}$ V   ___120/6 =_____ 2. $I = \underline{20}$ mA

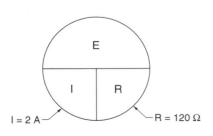

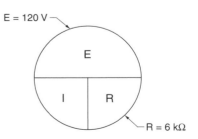

7

12 × 0.055   **3.** $P = .66$ mW   1200/2.5   **4.** $E = 480$ V

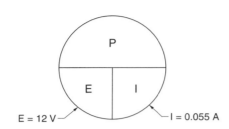
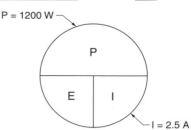

_____   **5.** $I = \_\_$ mA   _____   **6.** $R = \_\_$ Ω

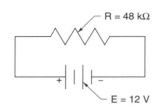
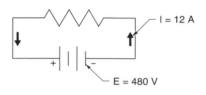

_____   **7.** $E = \_\_$ V   _____   **8.** $P = \_\_$ mW

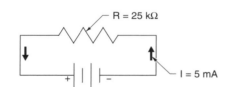
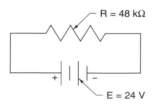

_____   **9.** $E = \_\_$ V   _____   **13.** $I = \_\_$ A
                  $I = 50$ mA $E = 24$ V
                  $R = 10$ kΩ $P = 24$ mW

_____   **10.** $R = \_\_$ Ω   _____   **14.** $E = \_\_$ V
                  $E = 26$ V $I = 3$ A
                  $I = 60$ mA $P = 45$ W

_____   **11.** $I = \_\_$ A   _____   **15.** $P = \_\_$ W
                  $E = 100$ V $E = 120$ V
                  $P = 100$ mW $I = 100$ mA

_____   **12.** $P = \_\_$ mW   _____   **16.** $I = \_\_$ A
                  $E = 12$ V $E = 480$ V
                  $I = 50$ mA $P = 1200$ W

_____ 17. R = ___ Ω
                    E = 220 V
                    I = 4000 mA

_____ 18. E = ___ mV
                    I = 15 μA
                    R = 15 kΩ

_____ 19. E = ___ V
                    I = 7 μA
                    R = 1 MΩ

_____ 20. E = ___ kV
                    I = 2 mA
                    R = 5 MΩ

_____ 21. I = ___ A
                    E = 480 V
                    P = 12 kW

_____ 22. P = ___ MW
                    E = 240 V
                    I = 125 A

_____ 23. I = ___ A
                    E = 36 V
                    P = 198 W

_____ 24. E = ___ V
                    I = 250 A
                    P = 50 kW

_____ 25. P = ___ W
                    E = 500 mV
                    I = 500 mA

_____ 26. I = ___ A
                    E = 208 V
                    P = 6240 W

Activity 1-3: Resistance in Series Circuits

Determine the total resistance in Series Circuit 1.

200+600+150= **1.** R_T = <u>950</u> Ω

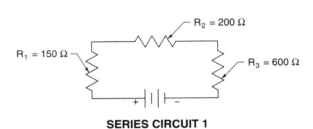

SERIES CIRCUIT 1

Determine the total resistance and unknown electrical quantity in Series Circuit 2.

100+500+600= **2.** R_T = <u>1200</u> Ω

120/1200 =   **3.** I_T = <u>0.1</u> mA

R_1 = 100 Ω, R_2 = 500 Ω, R_3 = 600 Ω, E = 120 V

SERIES CIRCUIT 2

Determine the total resistance in Series Circuit 3.

_____ 4. $R_T =$ ___ Ω

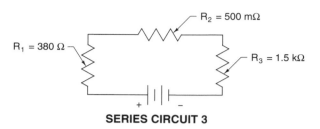

Determine the total resistance and unknown electrical quantities in Series Circuit 4.

_____ 5. $R_T =$ ___ kΩ

_____ 6. $I_T =$ ___ mA

_____ 7. $P_T =$ ___ W

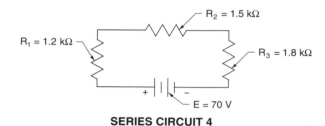

Activity 1-4: Resistance in Parallel Circuits

Determine the total resistance in Parallel Circuit 1.

_____ 1. $R_T =$ ___ Ω

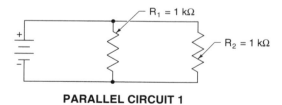

Determine the total resistance and unknown electrical quantity in Parallel Circuit 2.

_____ 2. $R_T =$ ___ Ω

_____ 3. $I_T =$ ___ mA

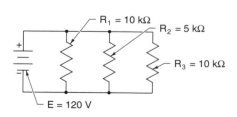

Determine the total resistance in Parallel Circuit 3.

_____ 4. $R_T =$ ___ Ω

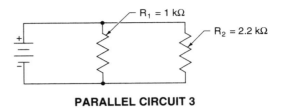

PARALLEL CIRCUIT 3

Determine the total resistance and unknown electrical quantity in Parallel Circuit 4.

_____ 5. $R_T =$ ___ Ω

_____ 6. $I_T =$ ___ mA

_____ 7. $P_T =$ ___ W

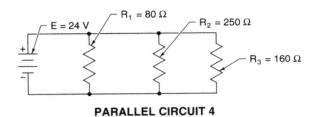

PARALLEL CIRCUIT 4

Activity 1-5: Resistance in Series/Parallel Circuits

Determine the total resistance in Series/Parallel Circuit 1.

_____ 1. $R_T =$ ___ Ω

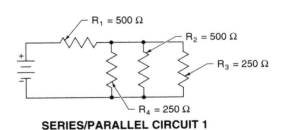

SERIES/PARALLEL CIRCUIT 1

Determine the total resistance in Series/Parallel Circuit 2.

_____ 2. $R_T =$ ___ Ω

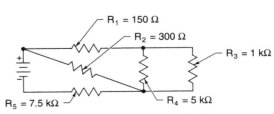

SERIES/PARALLEL CIRCUIT 2

Activity 1-6: Resistor Color Coding

State the resistance, tolerance, and resistance in simplest form using Resistor Color Codes on page 5.

_____ 1. Resistance is ___ Ω.

_____ 2. Tolerance is ___%.

_____ 3. The simplest form is ___.

_____ 4. Resistance is ___ Ω.

_____ 5. Tolerance is ___%.

_____ 6. The simplest form is ___.

_____ 7. Resistance is ___ Ω.

_____ 8. Tolerance is ___%.

_____ 9. The simplest form is ___.

_____ 10. Resistance is ___ Ω.

_____ 11. Tolerance is ___%.

_____ 12. The simplest form is ___.

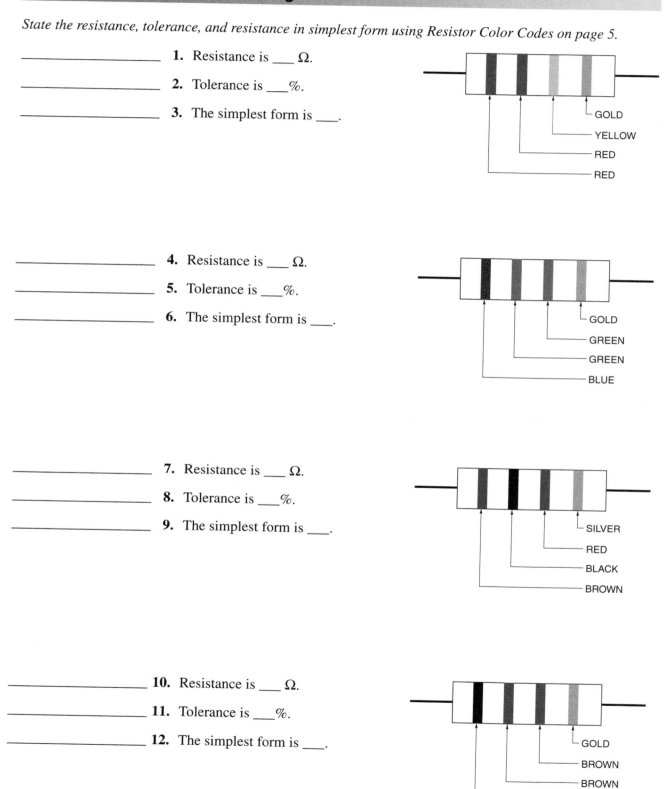

_____ 13. Resistance is ___ Ω.
_____ 14. Tolerance is ___%.
_____ 15. The simplest form is ___.

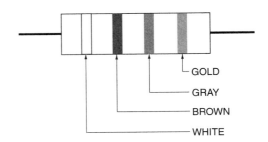

_____ 16. Resistance is ___ Ω.
_____ 17. Tolerance is ___%.
_____ 18. The simplest form is ___.

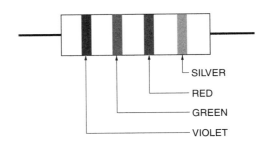

Activity 1-7: Finding Total Resistance of Resistors Connected in Parallel Using a Calculator

Use a calculator to find the total resistance of each circuit.

_____ 1. R_T = ___ Ω

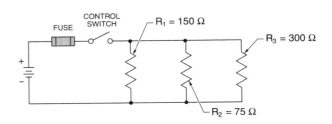

_____ 2. R_T = ___ Ω

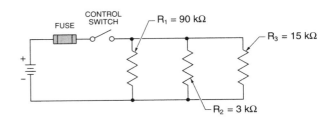

_____ 3. $R_T =$ ___ Ω

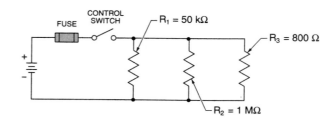

_____ 4. $R_T =$ ___ Ω

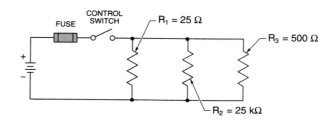

_____ 5. $R_T =$ ___ Ω

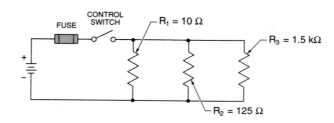

_____ 6. $R_T =$ ___ Ω

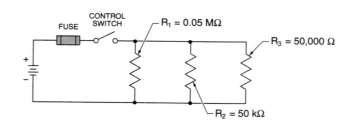

Electrical Tools and Test Instruments

Applications 2

Application 2-1: Multimeter Use

Multimeters

Electrical measurements are required when installing, operating, or repairing electrical equipment. A multimeter is the most common meter used to take electrical measurements. A *multimeter* is a meter that is capable of measuring two or more electrical quantities. Multimeters can be used to measure electrical functions such as voltage, current, continuity, resistance, capacitance, frequency, and duty cycle. Multimeters may be analog or digital.

Analog Multimeters

An *analog multimeter* is a meter that can measure two or more electrical quantities and displays the measured quantities along calibrated scales using a pointer. Analog multimeters use electromechanical components to display measured values. Most analog multimeters have several calibrated scales which correspond to the different selector switch settings (AC, DC, and R) and placement of the test leads (mA jack and 10 A jack). When reading a measurement on an analog multimeter, the function and range switches must be set at the correct quantity, and the correct scale must be read. **See Analog Multimeter.**

The function switch is set on AC when alternating voltage (VAC) is measured, and is set on DC when direct voltage or current (DC) is measured. If an analog multimeter does not have a separate setting for resistance, the positive DC setting is used. **See Function Switch.**

The normal setting for measuring DC voltage or current is +DC. This setting makes the red (+ marked) lead positive. The alternative setting for measuring DC voltage or current is –DC. This setting makes the red (+ marked) lead negative. This setting is used when it is preferable to have the test lead with the alligator clip (black) positive and the test lead with the pointer (red) negative.

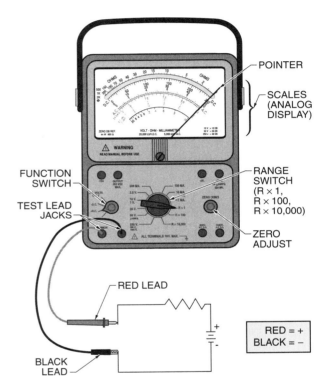

ANALOG MULTIMETER

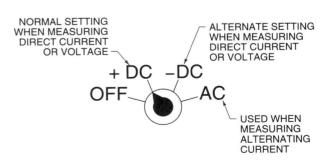

FUNCTION SWITCH

Analog Displays

An *analog display* is an electromechanical device that indicates readings by the mechanical motion of a pointer. Analog displays use scales to display measured values. Most analog multimeters have separate scales for reading voltage, current, and resistance. Analog scales may be linear or nonlinear. A *linear scale* is a scale that is divided into equally spaced segments. A *nonlinear scale* is a scale that is divided into unequally spaced segments. **See Analog Displays.**

Once the correct scale is determined, the number of units represented by the scale divisions must be determined. This value is then multiplied by the range switch setting to obtain the correct value. For example, if the pointer on an analog scale aligns with 4 and the range switch is set on R × 100, the value being measured is 400 Ω.

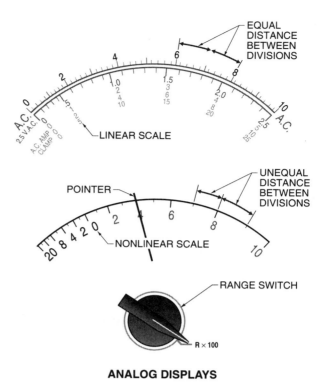

ANALOG DISPLAYS

Analog scales are divided using primary divisions, secondary divisions, and subdivisions. A *primary division* is a division with a listed value. A *secondary division* is a division that divides primary divisions in halves, thirds, fourths, fifths, etc. A *subdivision* is a division that divides secondary divisions in halves, thirds, fourths, fifths, etc. Secondary divisions and subdivisions do not have listed numerical values. When reading an analog scale, the primary, secondary, and subdivision readings are added to obtain the reading. **See Reading Analog Scales.**

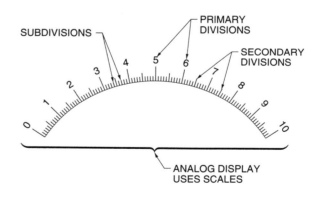

		READING
❶	PRIMARY DIVISION	2.0
❷	SECONDARY DIVISION	0.5
❸	SUBDIVISION	0.3
	METER READING	2.8

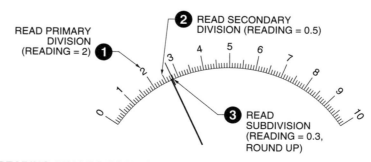

READING ANALOG SCALES

Digital Multimeters

A *digital multimeter (DMM)* is a meter that can measure two or more electrical properties and displays the measured properties as numerical values. The main advantages of a digital multimeter over an analog multimeter are the ability of a digital multimeter to record measurements, and ease in reading the displayed values. **See Digital Multimeter.**

Basic digital multimeters measure voltage, current, and resistance. Advanced digital multimeters include special functions such as capacitance and temperature measurement. DMM advanced features are helpful when troubleshooting problems such as improper frequencies, overheated circuit neutral wires, intermittent problems, power loss because of blown fuses, excessive current levels from overloaded circuits, and improper resistance from damaged insulation or components.

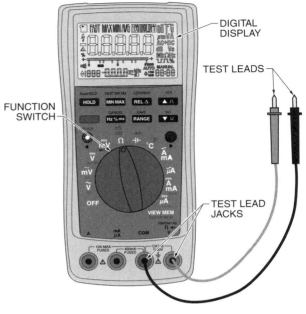

DIGITAL MULTIMETER

Reading Digital Displays

A *digital display* is an electronic device that displays readings on a meter as numerical values. Digital displays help eliminate human error when taking readings by displaying exact values measured. Errors occur when reading a digital display if the displayed prefixes, symbols, and/or decimal points are not properly applied. The exact value on a digital display is determined from the numbers displayed and the position of the decimal point. A range switch determines the placement of the decimal point. Accurate readings are obtained by using the range that gives the best resolution without overloading the meter. Most digital multimeters are autoranging, meaning that once a measurement function such as VAC is selected, the meter automatically selects the best meter range for taking the measurement. **See Digital Displays.**

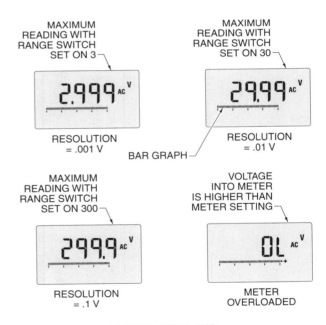

DIGITAL DISPLAYS

Bar Graphs

Most digital displays include a bar graph to show changes and trends in a circuit. A *bar graph* is a graph composed of segments that function as an analog pointer. The displayed bar graph segments increase as the measured value increases and decrease as the measured value decreases. The bar graph is used when quickly changing signals cause the digital display to flash or when there is a change in a circuit that is too rapid for the digital display to detect.

Application 2-2: Temperature Conversion

Converting Fahrenheit to Celsius

Converting between Fahrenheit and Celsius temperatures is often required in calculations and when using manufacturer's data. To convert a Fahrenheit temperature reading to Celsius, subtract 32 from the Fahrenheit reading and divide by 1.8. **See Fahrenheit Scale** and **Celsius Scale.** To convert Fahrenheit to Celsius, apply the following formula:

$$°C = \frac{(°F - 32)}{1.8}$$

where
°C = degrees Celsius
°F = degrees Fahrenheit
32 = difference between bases
1.8 = ratio between bases

Example: Converting Fahrenheit to Celsius
An insulated conductor is rated at 140°F. Convert the Fahrenheit temperature to Celsius.

$$°C = \frac{(°F - 32)}{1.8}$$
$$°C = \frac{(140 - 32)}{1.8}$$
$$°C = \frac{108}{1.8}$$
$$°C = \mathbf{60°C}$$

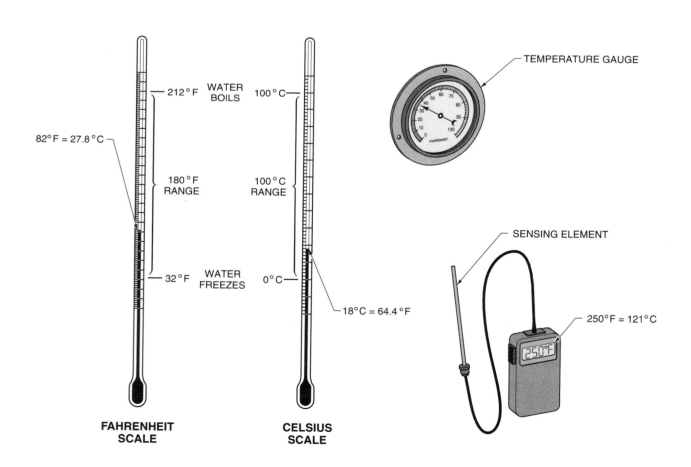

Converting Celsius to Fahrenheit

To convert a Celsius temperature reading to Fahrenheit, multiply 1.8 by the Celsius reading and add 32. To convert Celsius to Fahrenheit, apply the following formula:

$$°F = (1.8 \times °C) + 32$$

where
°F = degrees Fahrenheit
1.8 = ratio between bases
°C = degrees Celsius
32 = difference between bases

> ***Example: Converting Celsius to Fahrenheit***
> Convert 40°C to Fahrenheit.
> °F = (1.8 × °C) + 32
> °F = (1.8 × 40) + 32
> °F = 72 + 32
> °F = **104°F**

Application 2-3: Reading Graphs

Graphs

A *graph* is a diagram that shows the continuous relationship between two or more variables. Graphs present information in a simple form and are commonly used by component manufacturers to illustrate data and specifications. On a graph, one known variable is plotted horizontally and another is plotted vertically. The relationship between the two variables is represented by a straight or curved line. The point at which either variable line intersects the straight or curved line represents the value of the unknown variable.

A graph may be used to illustrate the effect of the ambient temperature on the operating characteristics of a fuse. *Ambient temperature* is the temperature of the air surrounding a fuse. As the ambient temperature increases, the opening time and capacity rating of the fuse decrease. As the ambient temperature decreases, the opening time and capacity rating of the fuse increase. For example, at 140°F, the opening time of a fuse is 70% of the standard rated opening time. Likewise, at −40°F, the opening time of a fuse is 133% of the standard rated opening time. **See Fuse Graph.**

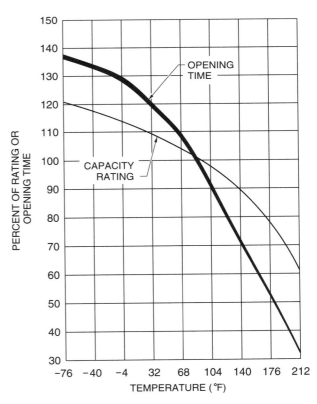

FUSE GRAPH

Plotting Graphs

Relationships between variables are often given in tables. Tables present the relationship between several variables. To find the continuous relationship between the variables, the table can be converted to a graph. A table is converted to a graph by applying the following procedure:

1. Draw two axes at right angles.
2. Label the two axes.
3. Select appropriate scales based on the values given.
4. Plot the points.
5. Draw a curve through the points.

Example: Plotting a Graph
The starting current draw of a motor varies with the speed of the motor. Plot the graph of the relationship. **See Motor Full-Load Current vs. Motor Speed Graph.**

1. Draw two axes at right angles.
2. Label the two axes.

Label the vertical axis "motor full-load current" and the horizontal axis "motor speed."

3. Select appropriate scales based on the values given.

Select 0 to 100 for the motor speed scale and 0 to 600 for the motor full-load current scale.

4. Plot the points.
5. Draw a curve through the points.

MOTOR SPEED (%)	100	98	95	87	72	0
MOTOR FLC (%)	100	200	300	400	500	600

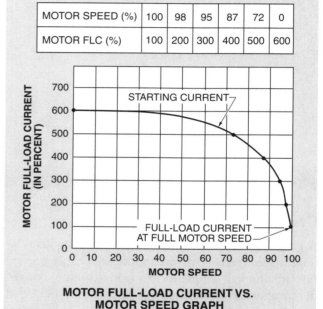

MOTOR FULL-LOAD CURRENT VS. MOTOR SPEED GRAPH

Application 2-4: Applying Electrical Principles when Troubleshooting

Troubleshooting Principles

Electrical meters are used to take measurements to verify that a circuit or component is working properly. Electrical meters are also used to troubleshoot a circuit or component that is not working properly. Before an electrical meter is connected into a circuit, the electrical quantity (voltage, current, resistance, etc.) must be understood in order to properly set the meter to take the measurement. For example, a meter set to measure a DC voltage cannot measure an AC voltage. Once the meter is set and connected into a circuit or to a component, series, parallel, and series/parallel circuits must be understood if the measured electrical quantity is to have any meaning. For example, when taking voltage measurements, the troubleshooter must understand that in a series circuit, the total applied voltage is divided across each load in the circuit, and in a parallel circuit, the total applied voltage is the same across each load in the circuit.

Electrical Tools and Test Instruments

Activities 2

Name _____ Date _____

Activity 2-1: Multimeter Use

List the correct reading for the function/range settings.

_____ 1. Reading = ___ mA (DC)   _____ 2. Reading = ___ μA (DC)

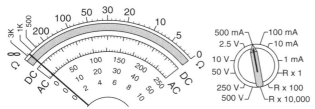 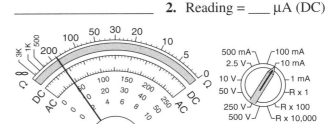

_____ 3. Reading = ___ kΩ   _____ 4. Reading = ___ VAC

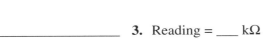 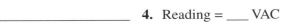

_____ 5. Reading = ___ VDC   _____ 6. Reading = ___ mA (DC)

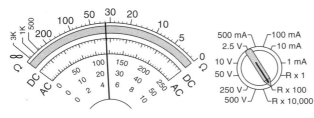 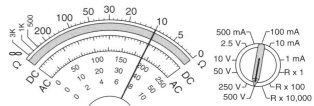

_____ 7. Reading = ___ VDC   _____ 8. Reading = ___ mA (DC)

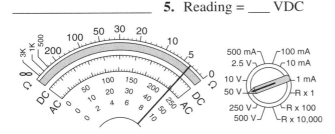 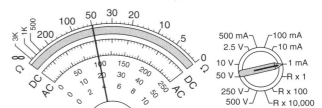

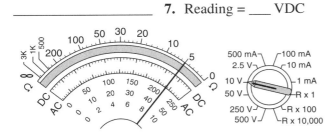 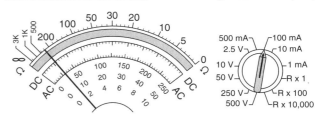

Activity 2-2: Temperature Conversion

Convert each Fahrenheit temperature reading to Celsius and each Celsius temperature reading to Fahrenheit.

_____ 1. 20°C = ___°F

_____ 2. An ambient temperature of 40°C is ___°F.

_____ 3. 460°C = ___°F

_____ 4. 140°F = ___°C

_____ 5. No. 8 THHN copper conductor can be used in temperatures below ___°F (186°C).

_____ 6. A temperature of 77°F is ___°C.

_____ 7. 1998°F = ___°C

_____ 8. The maximum operating temperature of FFH-2 fixture wire is ___°C (167°F).

_____ 9. 1610°C = ___°F

_____ 10. A temperature of 32°F is ___°C.

Activity 2-3: Reading Graphs

Complete the statements and answer the questions using Fuse Graph on page 19.

_____ 1. If a fuse is installed in an ambient temperature of 140°F, its current-carrying capacity is decreased by ___%.

_____ 2. If a fuse is installed in an ambient temperature of 140°F, its opening time is decreased by ___%.

_____ 3. If a 10 A rated fuse is installed in an ambient temperature of 32°F, the fuse carries a(n) ___ A load before opening.

_____ 4. Would a 10 A rated fuse installed in an ambient temperature of 32°F respond to an overcurrent faster or slower than its rated opening time?

_____ 5. If a 10 A rated fuse is installed in an ambient temperature of 176°F, the fuse carries a(n) ___ A load before opening.

_____ 6. Would a 10 A fuse installed in an ambient temperature of 176°F respond to an overcurrent faster or slower than its rated opening time?

Complete the tables and plot the relationships on the graphs.

7.

R =(Ω)	10	10	10	10	10	10	10	10	10	10
E =(V)	10	20	30	40	50	60	70	80	90	100
I =(A)										

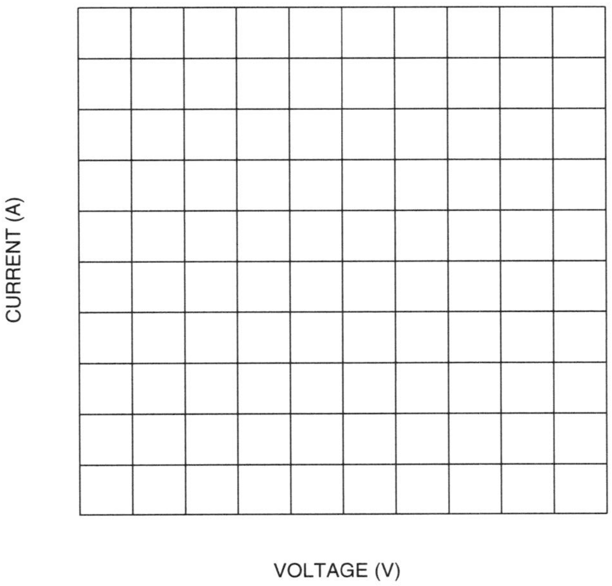

CURRENT (A)

VOLTAGE (V)

8.

R =(Ω)	100	100	100	100	100	100	100	100	100	100
E =(V)	2	4	6	8	10	12	14	16	18	20
I =(A)										

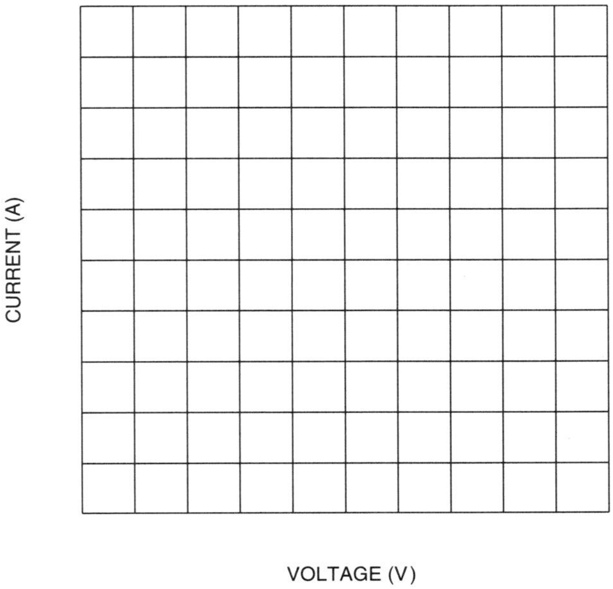

CURRENT (A)

VOLTAGE (V)

Activity 2-4: Applying Electrical Principles when Troubleshooting

1. Troubleshoot the heating element by setting the DMM to measure the resistance of each part of the heating element. Connect the test leads to measure the resistance of heating element 1 (R1).

_____ 2. If each heating element measures 30 Ω, what is the total resistance of the heating elements when all four are connected in series?

_____ 3. If each heating element measures 30 Ω, what is the total resistance of the heating elements when all four are connected in parallel?

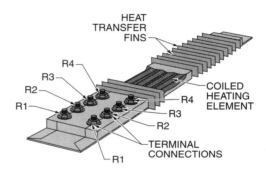

Electrical Safety

Applications 3

Application 3-1: Electrical Glove Selection

Electrical Gloves

Electricians wear electrical gloves to protect themselves from electrical shock. The three types of gloves an electrician may wear include insulated (rubber) gloves, cotton liners, and leather outer gloves. An insulated (rubber) glove (required) provides a high enough resistance to prevent electricity from entering the hand. A cotton liner (optional) is worn inside the insulated glove to add comfort and aid hand movement. A leather outer glove (highly recommended) is worn to protect the insulated glove.

An individual's glove size must be known in order to select a glove size that fits best and provides optimum comfort and protection. To determine glove size, the hand is held flat with the fingers together and thumb extended. **See Glove Sizing.** The circumference around the knuckles is measured and rounded up to the nearest ½″ (7, 7½, 8, etc.). The glove size is obtained by adding ½″ to the rounded measurement. A string may be used if a flexible tape measure is not available. The length of the string is measured with a standard tape measure or ruler.

Dielectric rubber gloves are typically seamless with a curved hand design for ease of use with hot and live electrical equipment. Gloves are available in solid black or in two-tone options (yellow/black or orange/black) that show color when the glove is worn through or damaged. Gloves are tested in accordance with ASTM D120 specifications.

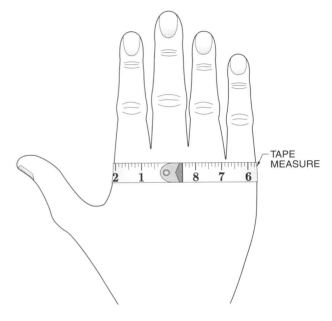

GLOVE SIZING

Example: Determining Glove Size
The circumference around the knuckles of an individual's hand is 8½″. Determine the glove size.
Glove size = length + ½″
Glove size = 8½″ + ½″
*Glove size = **9″***

25

Application 3-2: Grounding

Electrical Circuit Grounding

Grounding is the connection of all exposed non-current-carrying metal parts to the earth. Electrical circuits are grounded to safeguard equipment and personnel against the hazards of electrical shock. Voltage measurements are taken when checking a system for proper grounding. When checking to verify if a plug/receptacle is grounded, voltage measurements are taken from an ungrounded (hot) conductor to the ground conductor. If the full voltage is measured, the ground conductor is grounded. If the full voltage is not measured, or no voltage is measured, there is a grounding problem.

Plugs/receptacles have different configurations that indicate their voltage and current ratings. Plugs and receptacles are available in commercial, specification, and hospital grades. Commercial grade plugs are used in medium-duty applications and specification grade plugs are used in industrial applications. Hospital grade plugs are used in hospital, institutional, and industrial applications and are identified by a green dot on the face of the plug. **See Selected Nonlocking Wiring Devices. See Appendix.**

SELECTED NONLOCKING WIRING DEVICES

2-POLE, 3-WIRE

Wiring Diagram	NEMA ANSI	Receptacle Configuration	Rating
	5-15 C73.11		15 A 125 V
	5-20 C73.12		20 A 125 V
	5-30 C73.45		30 A 125 V
	7-15 C73.28		15 A 277 V

3-POLE, 4-WIRE

Wiring Diagram	NEMA ANSI	Receptacle Configuration	Rating
	14-15 C73.49		15 A 125/250 V
	14-20 C73.50		20 A 125/250 V

Application 3-3: Safety Label Information

Safety Labels

A *safety label* is a label that indicates areas or tasks that can pose a hazard to personnel and/or equipment. Safety labels use symbols and colors to designate the level of danger. Color is also used on conductors to indicate the conductor usage. The three most common signal words are danger, warning, and caution. **See Safety Labels.**

A danger signal word is used to indicate an imminently hazardous situation that could result in death or serious injury. The information indicated by the danger signal word indicates the most extreme type of potential situation, and must be followed. A warning signal word is used to indicate a potentially hazardous situation that could result in death or serious injury. A caution signal word is used to indicate a potentially hazardous situation that could result in minor or moderate injury. The information indicated by a caution signal word indicates a potential situation that may cause a problem to people and/or equipment.

SAFETY LABELS

Box Color	Symbol	Significance
red		**DANGER** – Indicates an imminently hazardous situation that, if not avoided, will result in death or serious injury
orange		**WARNING** – Indicates a potentially hazardous situation that, if not avoided, could result in death or serious injury
yellow		**CAUTION** – Indicates a potentially hazardous situation that, if not avoided, may result in minor or moderate injury, or damage to equipment. May also be used to alert against unsafe work practices
orange		**ELECTRICAL WARNING** – Indicates a high voltage location and conditions that could result in death or serious injury from an electrical shock
orange		**EXPLOSION WARNING** – Indicates location and conditions where exploding electrical parts may cause death or serious injury

Application 3-4: Troubleshooting Ground Problems

Ground Problems

When troubleshooting, approximate meter readings should be anticipated before a meter is connected into a circuit if the meter readings are going to have meaning and help determine circuit problems. Voltage measurements are taken to ensure nonconducting metal parts are grounded and the ground is properly connected. The three general categories of grounding are building grounding, equipment grounding, and electronic grounding. Building grounding ensures that a low-impedance (resistance) ground path for fault current (lightning) is present. Equipment grounding is used primarily to reduce the chance of electrical shock by ensuring all non-current-carrying metal parts are connected to ground. Electronic grounding is used to provide a clean chassis ground (usually less than 1 Ω) to help maintain signal integrity for sensitive electronic equipment. **See Grounding Systems.**

If a circuit or system is grounded, a voltage is present between a hot conductor (ungrounded energized conductor) and ground. On 120/240 V, 1φ, 3-wire systems and 120/203 V, 3φ, 4-wire systems, the voltage from a hot conductor to ground should be 120 VAC. On 277/480 V, 3φ, 4-wire systems, the voltage from a hot conductor to ground should be 277 VAC.

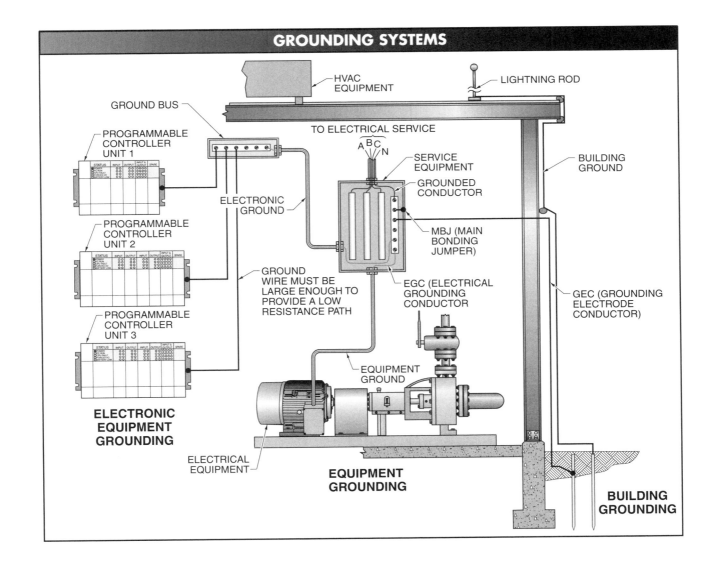

Electrical Safety

Name _____   **Date** _____

Activity 3-1: Electrical Glove Selection

Determine the proper glove size and model number.

_____ 1. The proper glove size for a hand with a circumference of 7¼" is ___.

_____ 2. A glove model number ___ is required for a hand with a circumference of 7¼" when working around 480 V or less.

_____ 3. The proper glove size for a hand with a circumference of 8¾" is ___.

_____ 4. A glove model number ___ is required for a hand with a circumference of 8¾" when working around 1500 V or less.

_____ 5. The proper glove size for a hand with a circumference of 9⅓" is ___.

_____ 6. A glove model number ___ is required for a hand with a circumference of 9⅓" when working around 600 V or less.

_____ 7. The proper glove size for a hand with a circumference of 7" is ___.

_____ 8. A glove model number ___ is required for a hand with a circumference of 7" when working around 1200 V or less.

ELECTRICAL GLOVES

Model Number	Size	Protection (in V)	Model Number	Size	Protection (in V)
7-1000	7	1000	9.5-20k	9.5	20,000
7-20k	7	20,000	10-1000	10	1000
7.5-1000	7.5	1000	10-20k	10	20,000
7.5-20k	7.5	20,000	10.5-1000	10.5	1000
8-1000	8	1000	10.5-20k	10.5	20,000
8-20k	8	20,000	11-1000	11	1000
8.5-1000	8.5	1000	11-20k	11	20,000
8.5-20k	8.5	20,000	11.5-1000	11.5	1000
9-1000	9	1000	11.5-20k	11.5	20,000
9-20k	9	20,000	12-1000	12	1000
9.5-1000	9.5	1000	12-20k	12	20,000

Activity 3-2: Grounding

Draw the correct position of the meter selector switch for the measurement to be taken. Connect and label the meter leads as indicated.

1. Connect the black (common) meter lead to the ground prong on the 30 A, 125 V receptacle and the red meter lead to a hot receptacle prong so the meter displays a voltage indicating that the receptacle is grounded.

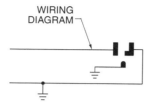
WIRING DIAGRAM

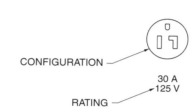
CONFIGURATION

RATING — 30 A, 125 V

2. Connect the black (common) meter lead to the ground prong on the 15 A, 277 V receptacle and the red meter lead to a hot receptacle prong so the meter displays a voltage indicating that the receptacle is grounded.

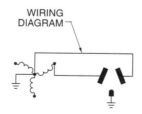
WIRING DIAGRAM

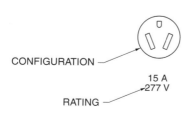
CONFIGURATION

RATING — 15 A, 277 V

3. Connect the black (common) meter lead to the ground prong on the 15 A, 125/250 V receptacle and the red meter lead to a hot receptacle prong so the meter displays a voltage indicating that the receptacle is grounded.

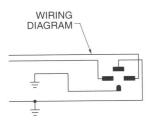

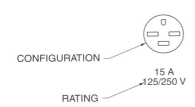

Activity 3-3: Safety Label Information

State the color of each safety label.

_____ 1. Color is ___.

_____ 2. Color is ___.

_____ 3. Color is ___.

_____ 4. Color is ___.

_____ 5. Color is ___.

Activity 3-4: Troubleshooting Ground Problems

Determine the expected DMM readings if the grounding system is properly wired.

_____ **1.** DMM 1 expected reading is ___ VAC.

_____ **2.** DMM 2 expected reading is ___ VAC.

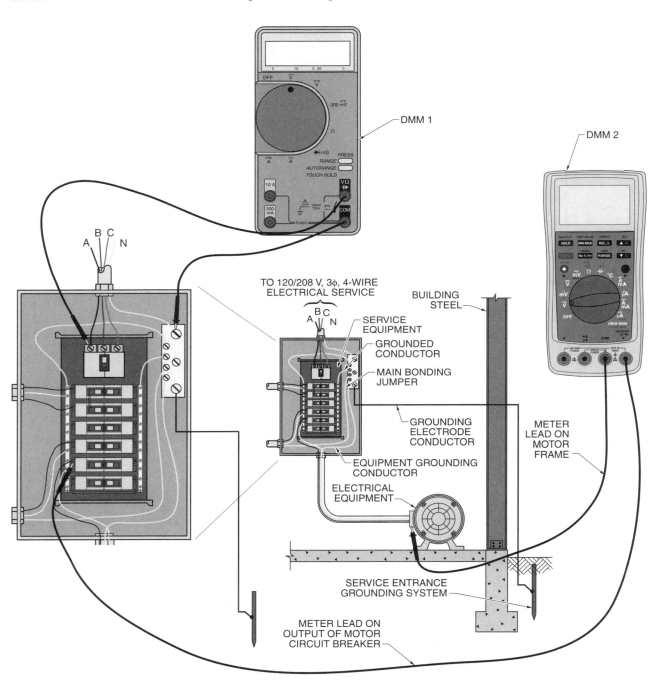

Electrical Symbols and Diagrams

Applications 4

Application 4-1: Symbol and Abbreviation Identification

Symbols and Abbreviations

Electrical prints are used when designing, troubleshooting, servicing, or repairing circuits. Electrical prints use standard symbols and abbreviations to show circuit operation and device use. A *symbol* is a graphic element that is used to conveniently represent electrical components in diagrams of most electrical and electronic circuits. An *abbreviation* is a letter or combination of letters that represents a word or phrase. **See Selected Symbols** and **Selected Abbreviations. See Industrial Electrical Symbols and Electrical/Electronic Abbreviations/Acronyms in Appendix.**

Symbols are used on electrical prints to identify the components used and how the components are interconnected. Each symbol must be understood in order to design, build, or troubleshoot an electrical circuit. Proper understanding of symbols is also necessary when ordering replacement components for new or existing electrical systems.

SELECTED SYMBOLS

Limit Switches		Pressure and Vacuum Switches	Temperature-Activated Switches	Timed Contacts Energized	Thermal Overload Relays	Control Transformers Single Voltage
NO	NC					
(NO symbol)	(NC symbol)	(symbol)	(symbol)	(symbol) NOTC	(symbol)	H1 H2 (transformer symbol)
HELD CLOSED	HELD OPEN	(symbol)	(symbol)	(symbol) NCTO		X2 X1

SELECTED ABBREVIATIONS

AC	ALTERNATING CURRENT	NO	NORMALLY OPEN
CB	CIRCUIT BREAKER	OL	OVERLOAD RELAY
CR	CONTROL RELAY	PB	PUSHBUTTON
DC	DIRECT CURRENT	PS	PRESSURE SWITCH
DP	DOUBLE POLE	R	REVERSE
DPST	DOUBLE POLE, SINGLE THROW	S	SWITCH
F	FORWARD	SOL	SOLENOID
LS	LIMIT SWITCH	SP	SINGLE POLE
M	MOTOR STARTER	SPDT	SINGLE POLE, DOUBLE THROW
MTR	MOTOR	SPST	SINGLE POLE, SINGLE THROW
NC	NORMALLY CLOSED	TR	TIME DELAY RELAY

Application 4-2: Wiring and Line Diagrams

Diagrams

A *wiring diagram* is a diagram that shows the placement and connections of all components in a control circuit or power circuit. A wiring diagram is used when troubleshooting, servicing, or repairing an operating circuit. When working with wiring diagrams, it is difficult to see the circuit operation because of the number of wires. To better understand the circuit operation, a line diagram is used. A *line diagram* is a diagram that shows the logic of an electrical circuit or system using standard symbols. A line diagram can be drawn from a wiring diagram by tracing each wire and drawing them in line diagram form.

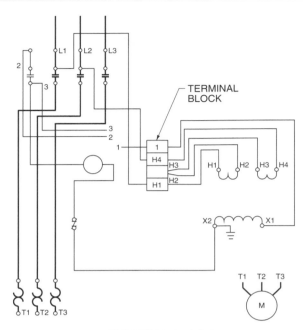

TERMINAL BLOCK

Terminal Block Wire Connections

Electrical connections are made using solder connections, clamp springs, and terminal blocks. A terminal block provides a convenient place to make wire changes and perform tests when troubleshooting. **See Terminal Block.**

Manufacturers provide tables that determine the maximum number of wires that can be connected to a terminal block. Terminal jumpers are used to connect several terminals together where additional terminal space is needed for additional wires in the same grouping (same numbers). Terminal strip manufacturers produce preassembled jumpers that enable efficient terminal connections to provide a professional appearance. Also, terminals are available in different colors to aid in keeping different parts of the circuit separated. It is always a good idea to include spare (unused) terminal points to allow expansion for future circuit additions and circuit modifications. **See Number of Wires per Terminal Block.**

Catalog Number	Wire Size (AWG)											
	#22	#20	#18	#16	#14	#12	#10	#8	#6	#4	#2	#1/0
	Number of the same size wires per terminal block											
001	3	3	3	2	1	—	—	—	—	—	—	—
002	4	3	3	3	2	1	—	—	—	—	—	—
003	4	4	4	3	2	2	1	1	—	—	—	—
004	—	4	4	3	2	2	1	1	—	—	—	—
005	5	5	4	4	3	2	2	1	1	—	—	—
006	—	—	—	—	—	4	4	4	3	2	1	1
007	4	4	4	3	2	2	1	—	—	—	—	—
008	4	4	3	2	2	1	—	—	—	—	—	—
009	4	4	4	3	3	2	1	—	—	—	—	—
010	6	5	5	4	3	2	1	—	—	—	—	—

Application 4-3: Allowable Voltage Drop across Conductors

Conductor Resistance

Electrical circuits and components are connected using conductors (wire). Conductor material includes copper, aluminum, copper-clad aluminum, steel, and bronze. Copper and aluminum are the most commonly used materials. Copper is the most common material used for most wiring, except for utility power distribution systems.

Since all conductors have resistance, all conductors have a voltage drop across them. The amount of voltage drop across a conductor reduces the supply voltage available for the loads. **See Lighting Circuit.**

If the amount of voltage drop across conductors is more than 3% of the supply voltage, the conductors are not correct for the application. The conductors may be undersized (wire too small), the run may be too long (increased total conductor resistance), or the amount of current drawn by the loads may be greater than the conductor rating. The resistance of a conductor is determined by the conductor material, size, and length. **See Copper Wire Specifications.**

The amount of voltage drop across a conductor increases if aluminum wire is used instead of copper (for same wire size), the wire length is increased, the wire size is decreased, or the load current is increased. The amount of voltage drop across a conductor can be calculated based on the amount of current the conductor is required to carry and the resistance of the conductor. To calculate conductor voltage drop, apply the following formulas:

$$I = \frac{P}{E}$$

where
I = maximum current of load/circuit (in A)
P = power requirement of load/circuit (in W)
E = voltage rating of load/circuit (in V)

$$E_D = I \times R$$

where
E_D = voltage drop of conductor (in V)
I = maximum current of load/circuit (in A)
R = resistance of conductor (in Ω)

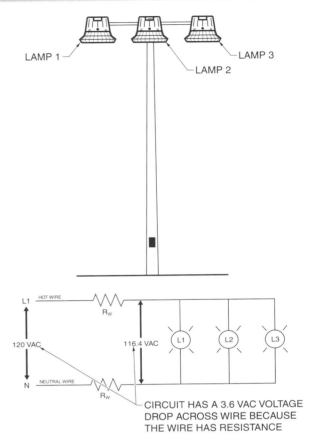

NOTE: R_W = RESISTANCE OF WIRE

LIGHTING CIRCUIT

COPPER WIRE SPECIFICATIONS			
AWG	Resistance*	Area†	Weight‡
18	6.51	1620	4.92
16	4.09	2580	7.82
14	2.58	4110	12.4
12	1.62	6530	19.8
10	1.02	10,400	31.4
8	0.641	16,500	50.0
6	0.403	26,300	79.5
4	0.253	41,700	126
2	0.159	66,400	201
0	0.100	106,000	319
0	0.079	133,000	403
0	0.063	168,000	508
0	0.050	212,000	641

* in Ω/1000′ at 77°F
† in circular mils
‡ in lb/1000′

$$E_{D\%} = \frac{E_D}{E} \times 100$$

where
$E_{D\%}$ = percent voltage drop of conductor
E_D = voltage drop of conductor (in V)
E = voltage rating of load/circuit (in V)

Example: Determining Percent Voltage Drop—230 VAC Circuit

What is the percent voltage drop across 1500′ of AWG #6 copper wire supplying power to a 2000 W, 230 VAC rated load?

1. Determine maximum current of circuit.

$$I = \frac{P}{E}$$
$$I = \frac{2000}{230}$$
$$I = 8.7 \text{ A}$$

2. Determine voltage drop of conductor. *Note:* A 1500′ length of #6 copper wire has a resistance of 0.604 Ω (0.000403 Ω/ft × 1500′ = 0.604 Ω).

$$E_D = I \times R$$
$$E_D = 8.7 \times 0.604$$
$$E_D = 5.25 \text{ V}$$

3. Determine percent voltage drop.

$$E_{D\%} = \frac{E_D}{E} \times 100$$
$$E_{D\%} = \frac{5.25}{230} \times 100$$
$$E_{D\%} = 0.0228 \times 100$$
$$E_{D\%} = 2.28\%$$

A voltage drop of 2.28% is less than 3% of the supply voltage, so the conductors are correct for the application.

Example: Determining Percent Voltage Drop—115 VAC Circuit

What is the percent voltage drop across 1500′ of AWG #6 copper wire supplying power to a 2000 W, 115 VAC rated load?

1. Determine maximum current of circuit.

$$I = \frac{P}{E}$$
$$I = \frac{2000}{115}$$
$$I = 17.4 \text{ A}$$

2. Determine voltage drop of conductor. *Note:* Conductor resistance equals 0.403 Ω per 1000′ of #6 copper wire, which equals 0.000403 Ω/ft (0.403 ÷ 1000 = 0.000403 Ω/ft). A 1500′ length of #6 copper wire has a resistance of 0.604 Ω (0.000403 Ω/ft × 1500′ = 0.604 Ω).

$$E_D = I \times R$$
$$E_D = 17.4 \times 0.604$$
$$E_D = 10.5 \text{ V}$$

3. Determine percent voltage drop.

$$E_{D\%} = \frac{E_D}{E} \times 100$$
$$E_{D\%} = \frac{10.5}{115} \times 100$$
$$E_{D\%} = 0.0913 \times 100$$
$$E_{D\%} = 9.1\%$$

A voltage drop of 9.1% is excessively high so the conductor size must be increased, the conductor length shortened, or loads removed (to reduce current). If loads cannot be removed or the conductor length shortened, the voltage can be increased if the loads are rated for a higher voltage.

Minimum Conductor Size Requirement

A conductor that has less than a 3% voltage drop across it may still be undersized because the National Electrical Code® (NEC®) places limits on the minimum conductor size that can be used for general wiring. Refer to NEC® Article 310 for information on conductors for general wiring.

Manufacturers typically include a minimum wire size chart that can be used when wiring their loads (motors, heating elements, appliances, etc.). The manufacturer chart takes into consideration the NEC® requirements as well as the wire type, size, and length of run. For example, a ¼ HP motor (single-phase or three-phase) could be wired with an AWG #18 copper conductor for a 25' run without the conductor dropping more than 3%. However, since the NEC® requires a minimum copper conductor size of AWG #14, the manufacturer wiring chart lists #14 copper conductor as the minimum size for wiring a ¼ HP motor. **See Single-Phase Motor Minimum Copper Wire Sizes** and **Three-Phase Motor Minimum Copper Wire Sizes.**

SINGLE-PHASE MOTOR MINIMUM COPPER WIRE SIZES*

Motor HP	Conductor Length†									
	25		50		100		150		200	
	115 V	230 V	115 V	230 V	115 V	230 V	115 V	230 V	115 V	230 V
⅛	14	14	14	14	12	14	10	14	8	14
⅙	14	14	12	14	10	14	6	14	6	12
¼	14	14	10	14	8	14	6	12	4	10
⅓	14	14	10	14	8	14	6	12	4	10
½	12	14	8	14	6	12	4	10	3	8
¾	10	14	6	12	4	10	2	8	1	6
1	10	14	6	12	4	10	2	8	1	6
1½	8	14	6	12	3	8	1	6	1/0	6
2	8	14	4	10	2	8	1/0	6	2/0	4
3	6	14	3	8	1/0	6	2/0	4	4/0	3

* AWG size
† in ft

THREE-PHASE MOTOR MINIMUM COPPER WIRE SIZES*

Motor HP	Conductor Length†								
	25–50			100			150–200		
	200 V	230 V	460 V	200 V	230 V	460 V	200 V	230 V	460 V
⅛	14	14	14	14	14	14	14	14	14
⅙	14	14	14	14	14	14	14	14	14
¼	14	14	14	14	14	14	14	14	14
⅓	14	14	14	14	14	14	12	14	14
½	14	14	14	12	14	14	10	12	14
¾	14	14	14	12	14	14	10	10	14
1	14	14	14	12	12	14	8	10	14
1½	12	14	14	10	10	14	6	8	14
2	12	12	14	8	10	14	6	6	12
3	10	12	14	6	8	14	4	6	12

* AWG size
† in ft

Application 4-4: Troubleshooting Power and Control Circuits

Manufacturer Troubleshooting Recommendations

Most manufacturers supply troubleshooting recommendations and symptom diagnostic guides with their equipment. Manufacturer information is traditionally found in the operations and maintenance (operator's) manual. Updates to the manual may be distributed as service bulletins. Manufacturer service bulletins may suggest test measurements that can be taken on the circuit to help isolate a problem. Manufacturer troubleshooting recommendations usually start with the most common problems, such as a blown fuse, open switch, etc.

Manufacturer Recommended Test 1

To test the circuit's incoming power, connect a voltmeter to the high-voltage side (primary) of the transformer. The voltage should read within +5% to –10% of the motor rated voltage. If it does not, troubleshoot the main power feeding the circuit. If it does, move on to Recommended Test 2.

Manufacturer Recommended Test 2

To test the control circuit, connect a voltmeter to the low-voltage side (secondary) of the transformer. The voltage should read within +5% to –10% of the motor starter coil rated voltage. If it does not, the transformer is bad or not wired correctly. If it does, move on to Recommended Test 3.

Manufacturer Recommended Test 3

To test the control circuit overcurrent protection device, connect one voltmeter lead to the grounded side of the control transformer and the other meter lead to the output of the control circuit protection device. If there is no voltage out of the protection device, the protection device is bad (open). If voltage out of the protection device is within +5% to –10% of the motor starter coil rated voltage, the protection device is good and testing should continue on to Recommended Test 4.

Manufacturer Recommended Test 4

To test the motor starter OLs, connect one voltmeter lead to the ungrounded side of the transformer and the other meter lead to the side of the OL contact that is connected to the motor starter coil. The voltage should read within +5% to –10% of the motor starter coil rated voltage. If it does not, the OL contact is open. Reset the overload contact.

Electrical Symbols and Diagrams

Activity 4 Activities

Name _____ Date _____

Activity 4-1: Symbol and Abbreviation Identification

Identify the abbreviations using the Solid-State Wiring Diagram.

_____ 1. T
_____ 2. CB
_____ 3. FU
_____ 4. PB-NO
_____ 5. PB-NC
_____ 6. FS
_____ 7. LS
_____ 8. ALM

_____ 9. GRD
_____ 10. SOL
_____ 11. SPST
_____ 12. FLS
_____ 13. PS
_____ 14. DIO
_____ 15. MTR

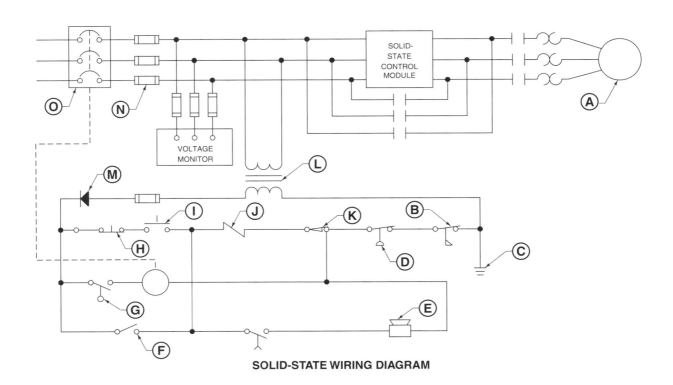

SOLID-STATE WIRING DIAGRAM

40 ELECTRICAL MOTOR CONTROLS *for Integrated Systems* APPLICATIONS MANUAL

Identify the symbols using the Starter Wiring Diagram.

_____ **16.** Normally open timed contacts

_____ **17.** Normally closed timed contacts

_____ **18.** Normally closed temperature switch

_____ **19.** Normally open pushbutton

_____ **20.** Normally closed pushbutton

_____ **21.** Normally open relay contacts

_____ **22.** Overload contact

_____ **23.** Motor starter

_____ **24.** Capacitor

_____ **25.** Ground

_____ **26.** Relay coil

_____ **27.** Timer coil

_____ **28.** Fuse

_____ **29.** Thermal overload

_____ **30.** Disconnect

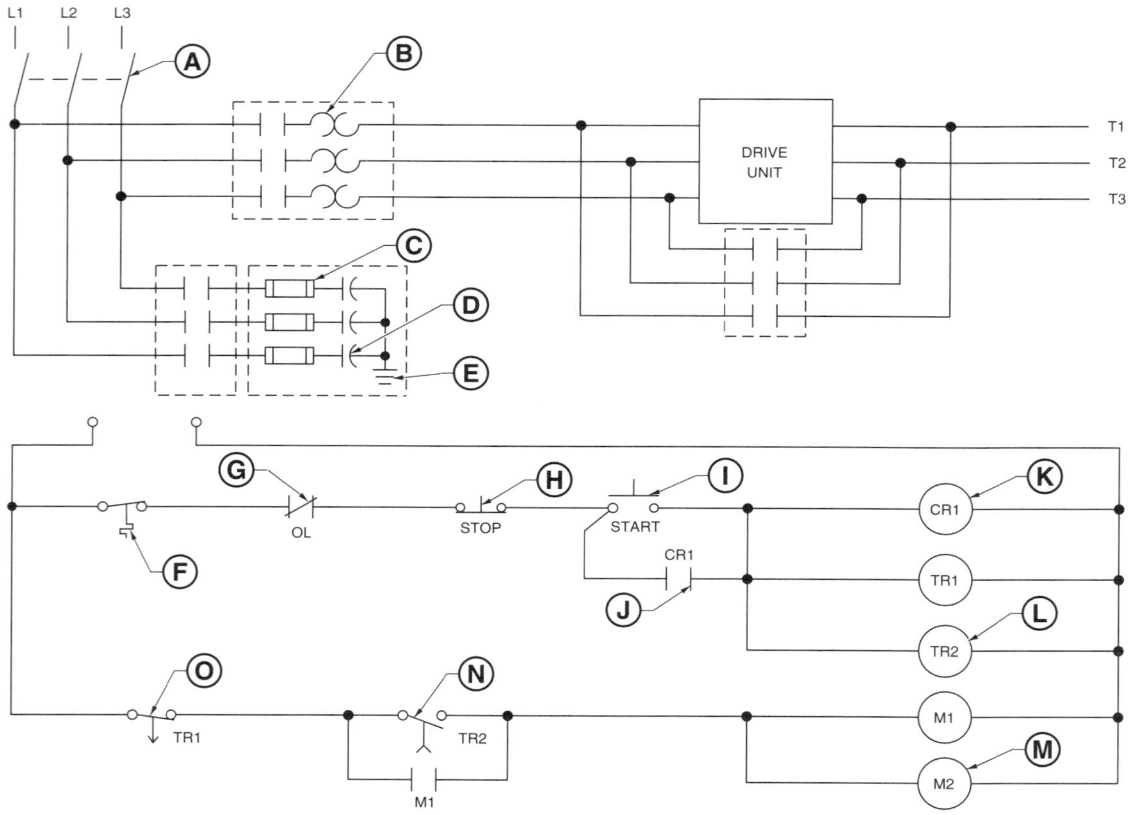

STARTER WIRING DIAGRAM

31. Add the electrical symbols to the Control Wiring Diagram.

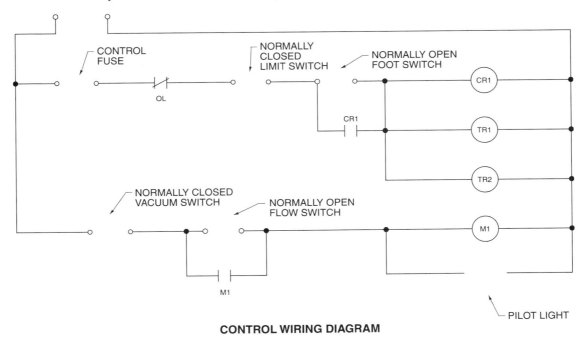

Identify each symbol on the Single-Phase-Input AC Motor Drive.

_____ **32.** Diode

_____ **33.** Coil

_____ **34.** Capacitor

_____ **35.** Transistor

_____ **36.** 3φ motor

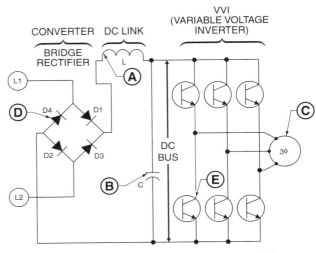

42 ELECTRICAL MOTOR CONTROLS *for Integrated Systems* APPLICATIONS MANUAL

Identify each symbol on the AC Primary Resistor Reduced-Voltage Starting Circuit.

_____ **37.** Transformer

_____ **38.** Resistor

_____ **39.** Normally open contact

_____ **40.** Fuse

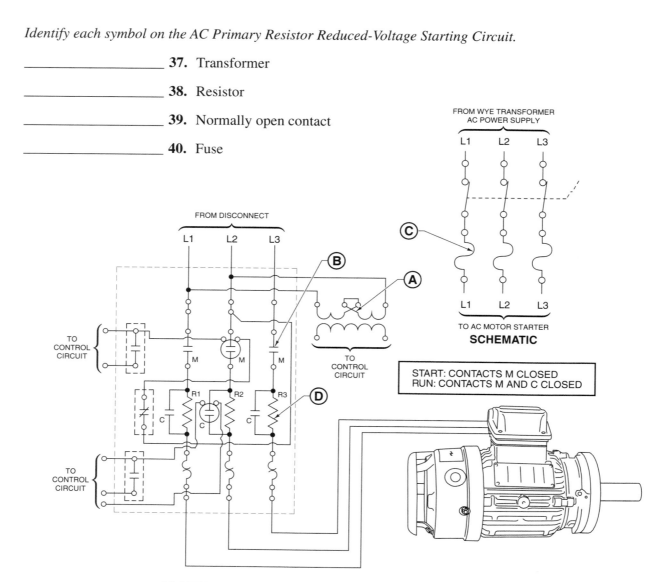

AC PRIMARY RESISTOR REDUCED-VOLTAGE STARTING CIRCUIT

Draw the symbol for each marked component in the Single-Voltage, Wye-Connected, Three-Phase Motor.

SINGLE-VOLTAGE, WYE-CONNECTED, THREE-PHASE MOTOR

Activity 4-2: Wiring and Line Diagrams

Answer the questions using the Number of Wires per Terminal Block chart on page 34.

_____ 1. The terminal block that can connect three #6 wires is number ___.

_____ 2. The terminal block that can connect six #22 wires is number ___.

_____ 3. The terminal block that can connect five #18 wires is number ___.

_____ 4. The maximum number of #14 wires that can be connected using a #001 terminal block is ___.

_____ 5. The maximum number of #14 wires that can be connected using a #007 terminal block is ___.

_____ 6. Can a pair of #2 wires be connected using a #006 terminal block?

Activity 4-3: Allowable Voltage Drop across Conductors

Calculate the voltage drop and percent voltage drop for each application.

_____ 1. Voltage drop is ___V.

_____ 2. Percent voltage drop is ___%.

_____ 3. Voltage drop is ___V.

_____ 4. Percent voltage drop is ___%.

31 A LOAD
ACTUAL SUPPLY = 117 VAC
TOTAL CONDUCTOR LENGTH = 150′
WIRE = COPPER #10

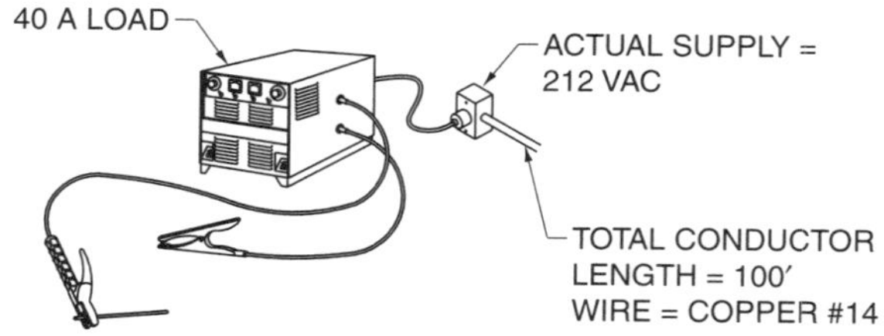

40 A LOAD
ACTUAL SUPPLY = 212 VAC
TOTAL CONDUCTOR LENGTH = 100′
WIRE = COPPER #14

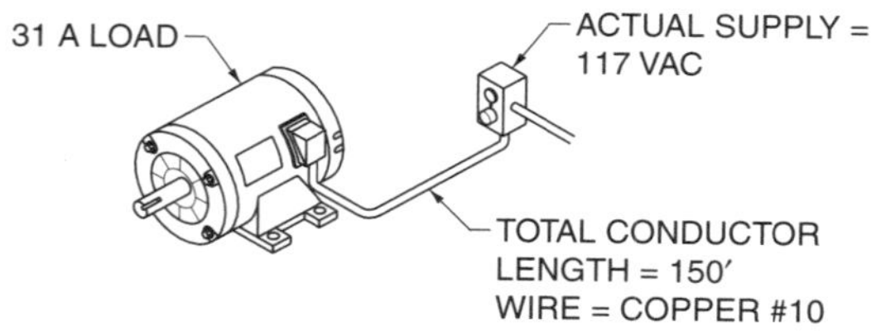

_____ 5. Voltage drop is ___V.

_____ 6. Percent voltage drop is ___%.

_____ 7. Voltage drop is ___V.

_____ 8. Percent voltage drop is ___%.

8 A LOAD
ACTUAL SUPPLY = 109 VAC
TOTAL CONDUCTOR LENGTH = 80′
WIRE = COPPER #12

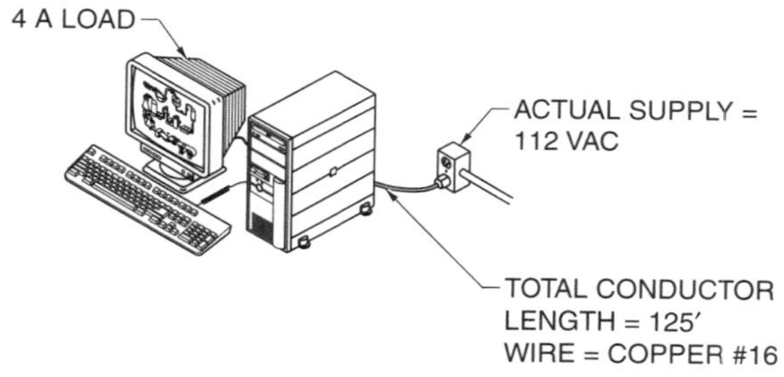

4 A LOAD
ACTUAL SUPPLY = 112 VAC
TOTAL CONDUCTOR LENGTH = 125′
WIRE = COPPER #16

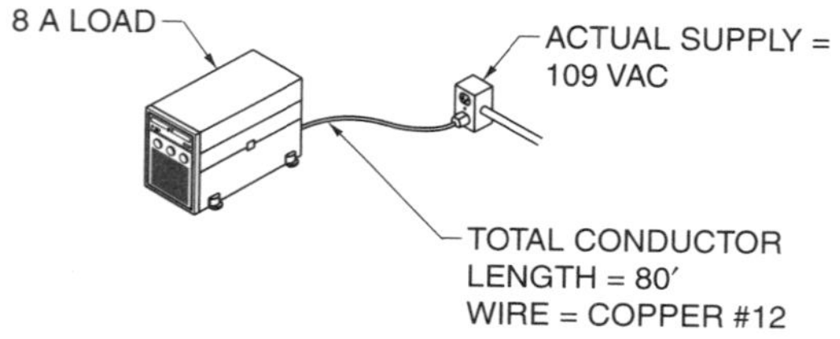

Activity 4-4: Troubleshooting Power and Control Circuits

1. Connect DMM 1 to the high-voltage side (primary) of the transformer to test the circuit incoming power. Connect DMM 2 to the low-voltage side (secondary) of the transformer to test the control circuit. Connect one lead of DMM 3 to the grounded side of the control transformer and the other lead to the output of the control circuit protection device to test the control circuit overcurrent protection device. Connect one lead of DMM 4 to the ungrounded side of the transformer and the other lead to the side of the OL contact that is connected to the motor starter coil to test the motor starter overloads.

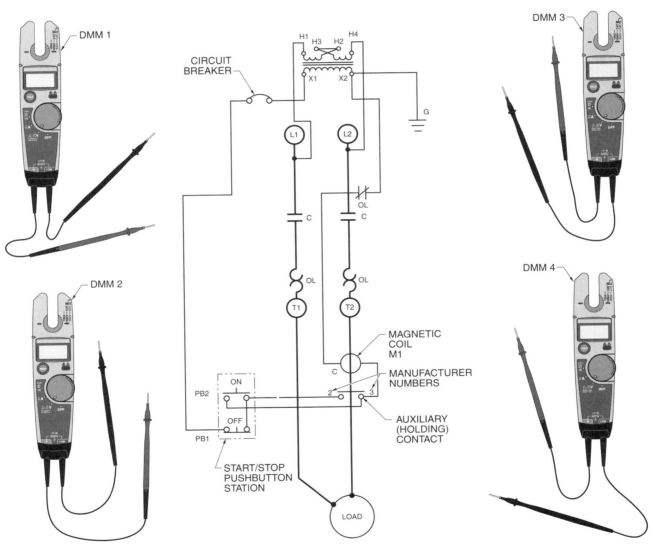

WIRING DIAGRAM

46 ELECTRICAL MOTOR CONTROLS *for Integrated Systems* APPLICATIONS MANUAL

When troubleshooting power and control circuits, approximate meter readings should be anticipated if the meter readings are going to be used to help determine circuit problems. Determine the expected DMM readings if the circuit is working properly.

_____ **2.** The expected reading of DMM 1 with the motor ON is ___ VAC.

_____ **3.** The expected reading of DMM 2 with the motor ON is ___ VAC.

_____ **4.** The expected reading of DMM 3 with the motor ON is ___ mA.

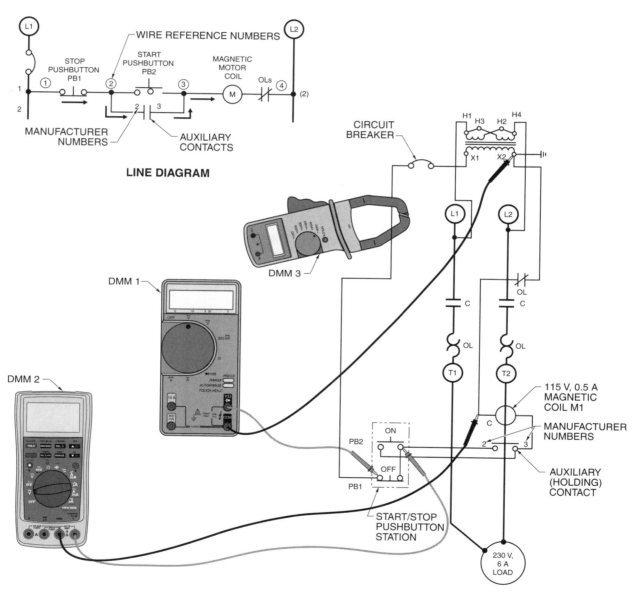

Logic Applied to Line Diagrams

5 Applications

Application 5-1: Numerical Cross-Reference Numbers

Line Diagrams

A line diagram (ladder diagram) is a diagram that shows the logic of a control circuit in its simplest form. A line diagram consists of a series of symbols interconnected by lines that are laid out like rungs on a ladder. A line diagram does not show the location of each component in relationship to the other components in the circuit. A line diagram is used when designing, modifying, or describing a circuit. The arrangement of a line diagram should promote clarity. Graphic symbols, abbreviations, and device designations are drawn per standards. The circuit should be shown in the most direct path and the most logical sequence. Lines between symbols can be horizontal or vertical, but should be drawn so as to minimize line crossing. Line diagrams provide a fast, easy understanding of the connections and use of components. **See Wiring Diagram** and **Line Diagram**.

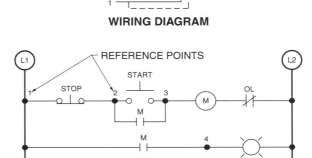

Line Number References

Each line in a line diagram should be numbered starting with the top line and reading downward. Numbering each line simplifies the understanding of the function of the circuit. The importance of this numbering system becomes clear as circuits become more complex and lines are added. **See Line Number References.**

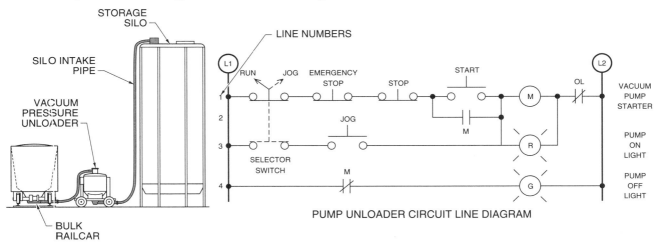

47

Numerical Cross-Reference System

Numerical cross-reference numbers are used to help identify where and how contacts are used. Relays, contactors, and magnetic motor starters normally have more than one set of auxiliary contacts. These contacts may appear at several different locations in a line diagram. A numerical cross-reference system quickly identifies the location and type of contacts controlled by a given device. The numerical cross-reference numbers are included in parentheses to the right of the line diagram. Normally open contacts are represented by line numbers. Normally closed contacts are represented by a line number that is underlined. The line numbers refer to the line on which the contacts are located. **See Numerical Cross-Reference System.**

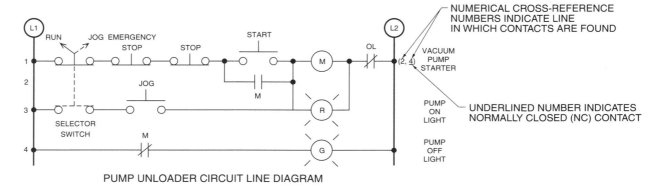

NUMERICAL CROSS-REFERENCE SYSTEM

Application 5-2: Assigning Wire-Reference (Terminal) Numbers

Wire-Reference Numbers

Each wire in a control circuit is assigned a reference point on a line diagram to keep track of the different wires that connect the components in the circuit. Each reference point is assigned a reference number. Reference numbers are normally assigned from the top left to the bottom right.

When assigning wire-reference numbers, any wire that is always connected to a point is assigned the same number. The wires that are assigned a number vary from two to the number required by the circuit. Any wire that is prewired when the component is purchased is normally not assigned a reference number. When assigning reference numbers, different numbering assignments can be used. The exact numbering system varies for each manufacturer or design engineer. This numbering system applies to any control circuit such as single station, multistation, reversing, or two-speed circuits.

For example, a line diagram may have three reference points. The reference points may be assigned the numbers 1, 2, and 3; L1, 1, and L2; 1, 2, and L2; or 1, 3, and 2. The first reference point is labeled 1 or L1. Any wire connected to this point at all times is labeled 1. Although the 1, 2, and 3 method has been the most common, numbering L1 as 1 and L2 as 2 is the most modern and accepted method because it always identifies the two power lines (L1 and L2) as terminals 1 and 2 in the control panel. **See Wire-Reference (Terminal) Numbers.**

Chapter 5—Logic Applied to Line Diagrams **49**

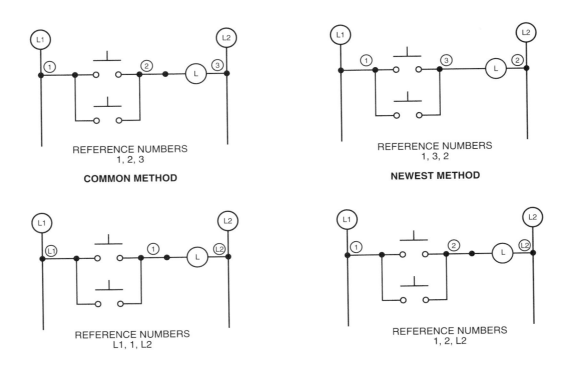

WIRE-REFERENCE (TERMINAL) NUMBERS

Application 5-3: Manufacturer Terminal Numbers

Terminal Numbers

Electrical equipment manufacturers include terminal numbers at the connection points for ease of circuit wiring and troubleshooting. In order to properly wire electrical components, the connections must be made to the correct terminals. Manufacturer terminal numbers are often added to a line diagram after the specific equipment to be used in the control circuit is identified. These terminal numbers are used to identify and separate the different components (coil, normally closed contacts, etc.) included on the individual pieces of equipment. **See Manufacturer Terminal Numbers.**

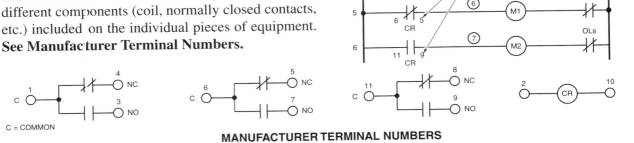

MANUFACTURER TERMINAL NUMBERS

Application 5-4: Wiring Control Panels

Wiring Control

Control panels are used to house the control parts of an electrical system that are not inputs (pushbuttons, limit switches, etc.) or outputs (motors, solenoids, etc.). Inputs such as pushbuttons or liquid level switches must be located at set locations. However, control devices such as motor starters, relays, and timers can be located in one centralized control panel. Control panels help organize electrical circuit wiring and aid in troubleshooting. **See Wiring Control Panels.**

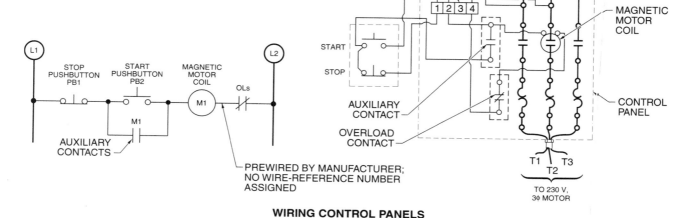

WIRING CONTROL PANELS

Application 5-5: Basic Switching Logic

Logic Functions

Control circuits are designed to perform a specific function. All control circuits are basic logic functions or combinations of logic functions. Logic functions include AND, OR, NOT, NOR, and NAND. The logic function depends on the relationship between the input and output signals of a circuit. Logic functions are used in all areas of industry including electricity, electronics, hydraulics, and pneumatics, as well as in math and other routine activities.

An input is the switch or switches that start or stop the flow of electricity to outputs. An output is the load or loads that use electricity to produce work. Typical loads include lights, motors, heating elements, and solenoids. A circuit is activated when the switch contacts are opened or closed manually, as by a pushbutton; mechanically, as by a limit switch; or automatically, as by a temperature switch. Control devices are connected into a circuit so that the circuit can function in a predetermined manner. **See Basic Logic Functions.**

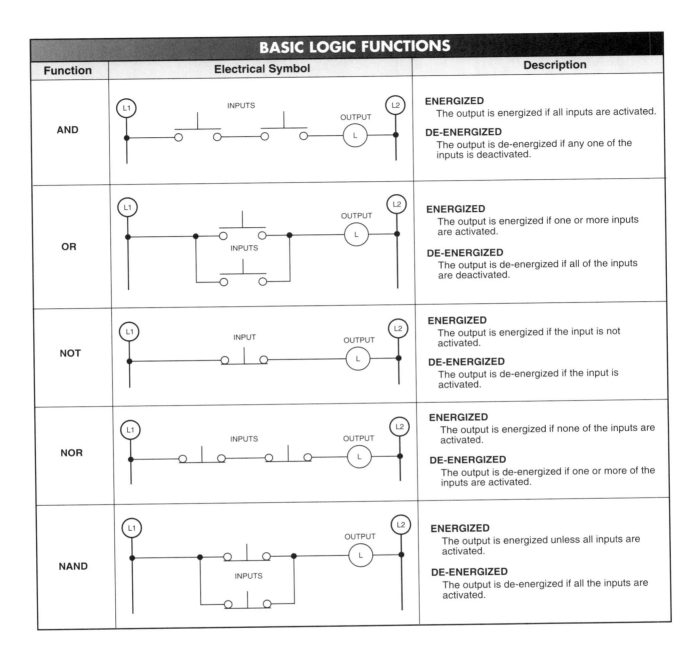

Application 5-6: Troubleshooting Control Circuits

Control Circuits

A control circuit is the part of a circuit that determines when and how the loads are turned ON and OFF. Troubleshooting a control circuit is a matter of finding the point where the control power is lost. The point where the control power is lost usually indicates a malfunctioning switch, interface, or load in the control circuit. Troubleshooting requires the use of electrical test equipment, drawings and diagrams, and manufacturer specifications. **See Control Circuit.**

52 ELECTRICAL MOTOR CONTROLS *for Integrated Systems* APPLICATIONS MANUAL

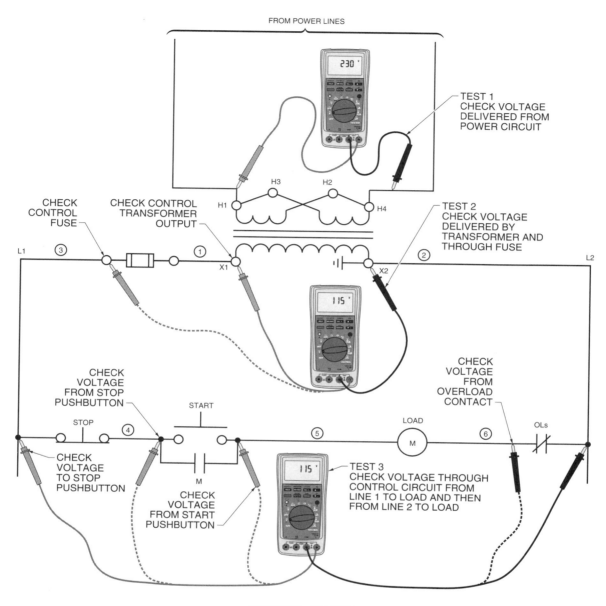

CONTROL CIRCUIT

To troubleshoot a control circuit, apply the following procedure:

Test 1. Check the voltage delivered from the power circuit. All control circuits receive voltage from a power circuit. Check the voltage coming from the power circuit to ensure the voltage is present and at the correct level. If the voltage from L1 to L2 is not correct, there is a problem in the power circuit delivering power to the control circuit.

Test 2. Check the voltage delivered through the control transformer. Most control circuits operate at a lower voltage level than the power circuit. A control transformer is used to reduce the voltage to the level required by the control circuit.

Most control transformers have a fuse on the secondary side, or a fuse is added on the primary side. Check that the correct voltage is delivered by the transformer and is delivered through the fuse. The primary (input side) of the transformer must have the proper voltage or the transformer cannot deliver the correct (secondary) voltage to operate the control circuit. The primary transformer voltage should be within +5% to –10% of the transformer input rating.

If the voltage into the transformer is within an acceptable range, check the output of the transformer. The output (secondary voltage) of the transformer should be within +5% to –10% of the transformer output rating. As part of checking the transformer output, also check the control circuit fuse since the fuse is usually located on or near the transformer. If there is a voltage into and out of the fuse, the fuse is good. If there is voltage into the fuse but not out of the fuse, the fuse is bad.

Test 3. Check the voltage through the control circuit. This is usually done by connecting the meter leads across the power supply (L1 and L2) and moving one meter lead at a time throughout the control circuit. Connect one side of a DMM set to measure voltage to L2. Move the other lead throughout the control circuit. Test voltage into and out of each device in the circuit.

If voltage is going into a control device such as a pushbutton or overload contact, and coming out of the device, the device is closed. If the voltage is going into a control device but not coming out, the device is open. When a device does not pass the voltage as required, replace or service that device.

Troubleshooting at Terminal Strips

When troubleshooting, approximate meter readings should be anticipated if the meter readings are going to have meaning and help determine circuit problems. Troubleshooting is made easier by first taking measurements at the terminal strip inside the control panel. **See Terminal Strips.**

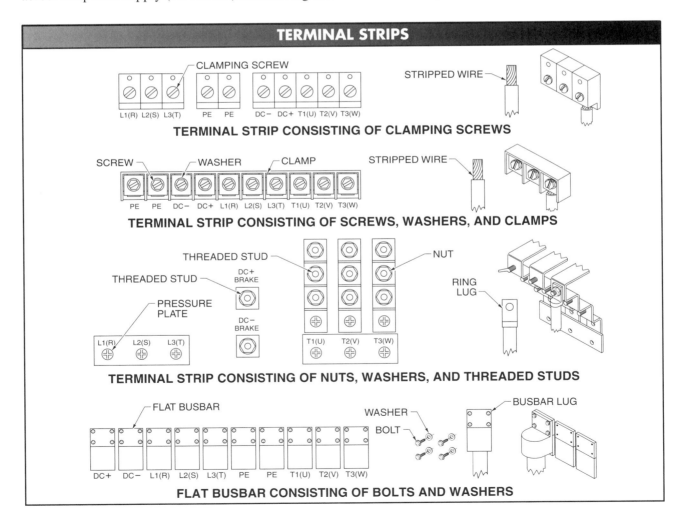

Troubleshooting at the terminal strip involves troubleshooting the control circuit, since the control circuit components are connected to the terminal strip. The control circuit includes all of the low-voltage (24 V, 115 V, etc.) control components such as pushbuttons, secondary of control transformer, motor starter coil, overload contacts, and any other components connected into the low-voltage circuit. Troubleshooting the control circuit involves first verifying that the control circuit is powered by checking secondary voltage from transformer, checking that the control circuit fuses are good, and then working through the control circuit by testing each component.

Wire-reference numbers and manufacturer terminal numbers are used when troubleshooting a circuit. In order to add meaning to the troubleshooting process, the circuit line diagram must be understood and used for the proper placement of the meter leads on the terminal strip inside the control panel. **See Troubleshooting at Terminal Strips.**

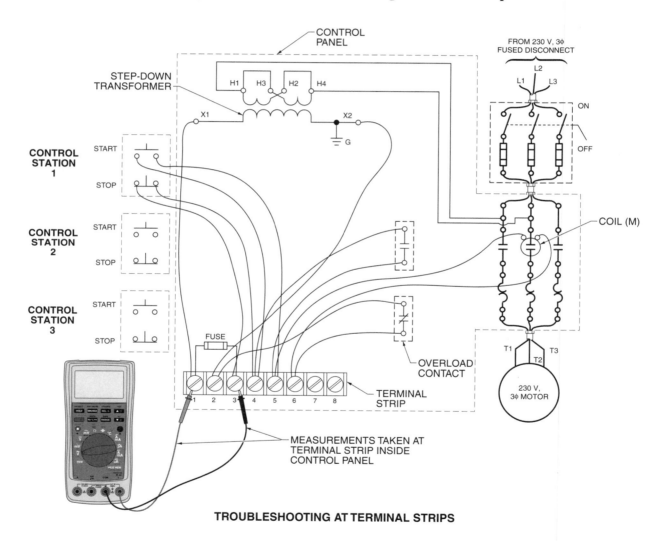

TROUBLESHOOTING AT TERMINAL STRIPS

Logic Applied to Line Diagrams

5 Activities

Name _____ Date _____

Activity 5-1: Numerical Cross-Reference Numbers

1. Add the numerical cross-reference numbers for the two control relays and timer in the Security Door Lock System line diagram.

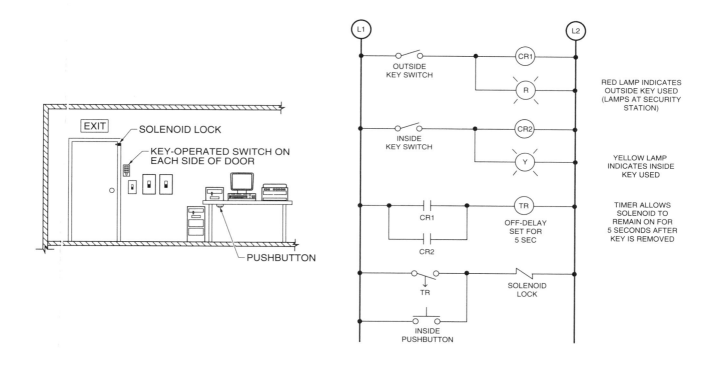

SECURITY DOOR LOCK SYSTEM

55

Activity 5-2: Assigning Wire-Reference (Terminal) Numbers

1. List the reference wire (terminal) number for each wire in the Security Door Lock System. Start by marking L1 wire number 1 and L2 wire number 2. After marking wires number 1 and 2, assign numbers from the top left of the diagram to the bottom right. Number every wire in the circuit.

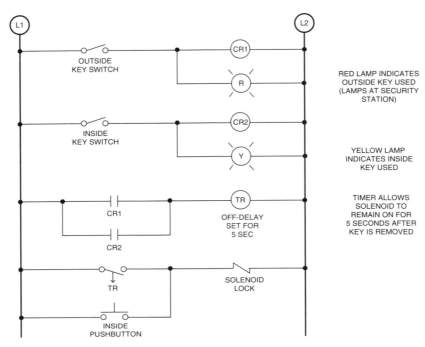

SECURITY DOOR LOCK SYSTEM

When wiring control stations, wire-reference numbers are used (usually added to wires) to help keep track of the wires in the system. Given each line diagram with wire-reference numbers, list the wire-reference number that would be used to identify each wire coming from the control stations.

_____ 2.
_____ 3.
_____ 4.
_____ 5.

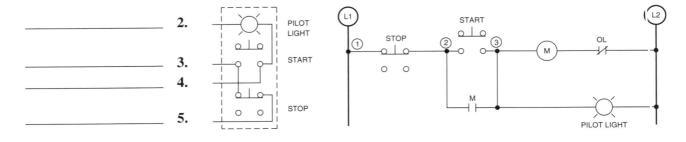

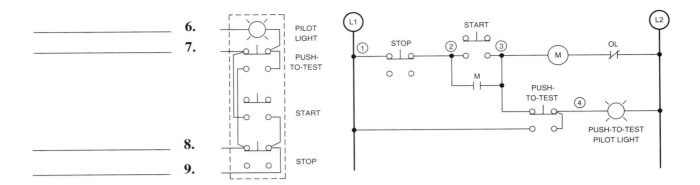

List the reference wire number for each wire on the line diagram and pushbutton station of the single station control circuits. Mark each wire except the wire connecting the starting coil to the overload contacts.

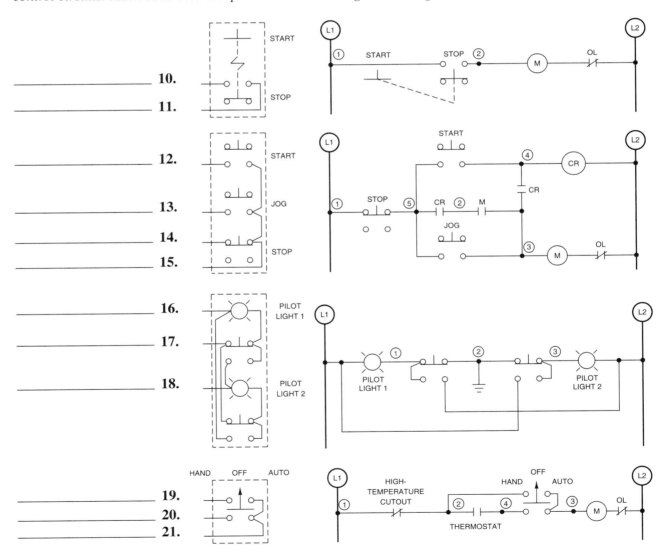

List the reference wire number for each wire on the line diagram and pushbutton station of the multistation control circuits. Mark each wire except the wire connecting the starting coil to the overload contacts.

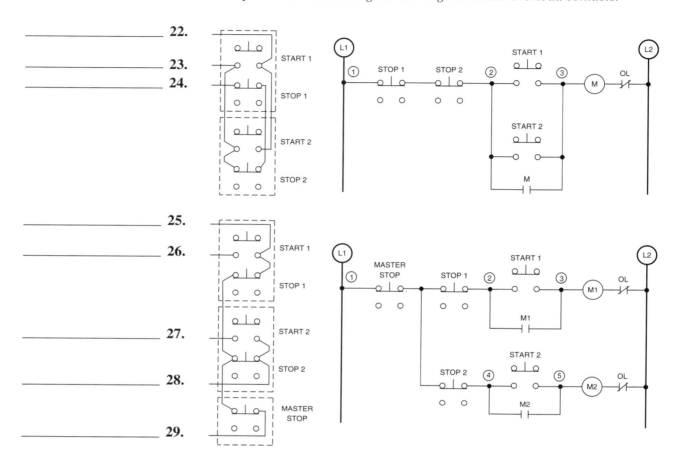

List the reference wire number for each wire on the line diagram and pushbutton station of the reversing control circuits. Mark each wire except the wire connecting the starting coil to the overload contacts.

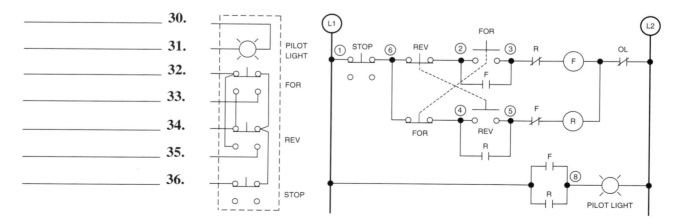

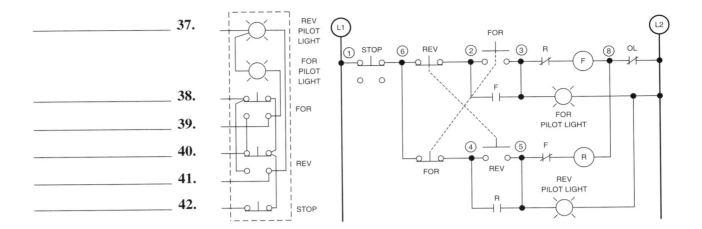

_____ 37.
_____ 38.
_____ 39.
_____ 40.
_____ 41.
_____ 42.

List the reference wire number for each wire on the line diagram and pushbutton station of the two-speed control circuit. Mark each wire except the wire connecting the starting coil to the overload contacts.

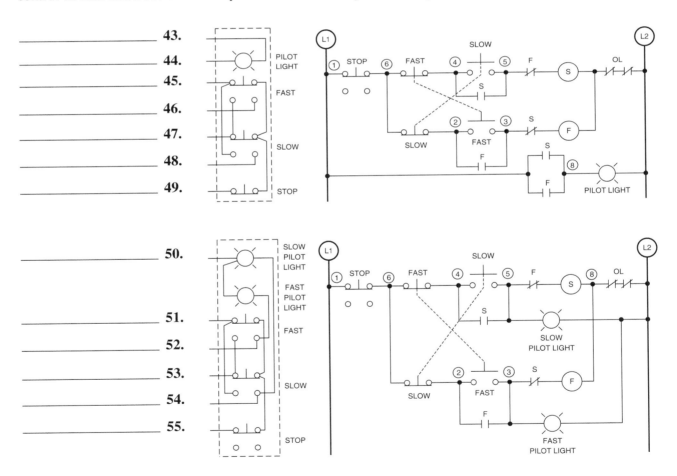

_____ 43.
_____ 44.
_____ 45.
_____ 46.
_____ 47.
_____ 48.
_____ 49.

_____ 50.
_____ 51.
_____ 52.
_____ 53.
_____ 54.
_____ 55.

Activity 5-3: Manufacturer Terminal Numbers

1. For each control relay and the timer, identify the manufacturer terminals that could be used when wiring the Security Door Lock System circuit. In this circuit, three different manufacturers' relay/timer equipment is used. With some equipment, there may be more than one set of contacts that can be used. List both sets of contacts that can be used.

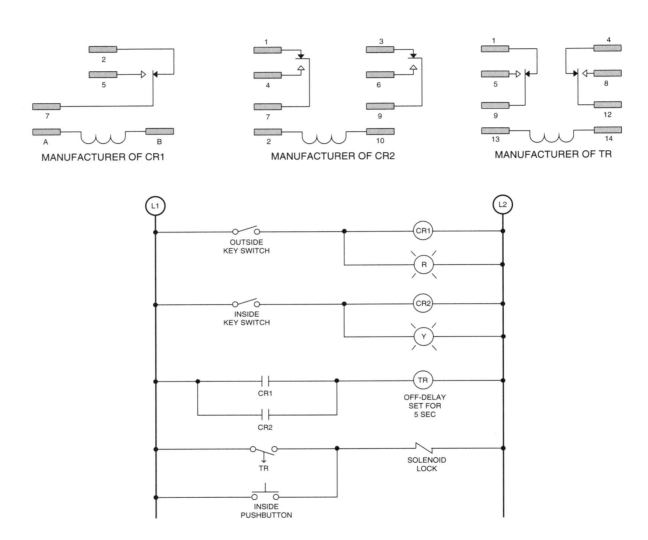

SECURITY DOOR LOCK SYSTEM

Activity 5-4: Wiring Control Panels

1. Using the wire-reference numbers (terminal numbers) from the Security Door Lock System, connect (wire) each device to the terminal strip in the control panel.

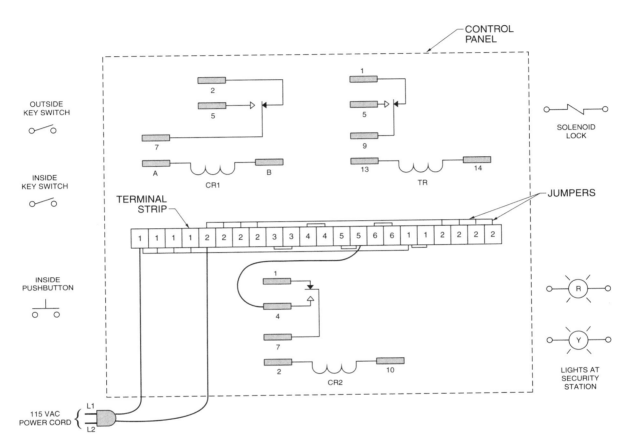

SECURITY DOOR LOCK SYSTEM

Activity 5-5: Basic Switching Logic

Identify the basic logic function of each circuit.

_____ 1. In the Product Stamp Circuit, PB1, PB2, the foot switch, and the pressure switch are connected in ___ circuit logic.

_____ 2. In the Product Stamp Circuit, the key operated switch is connected in ___ circuit logic with PB1, PB2, the foot switch, and the pressure switch.

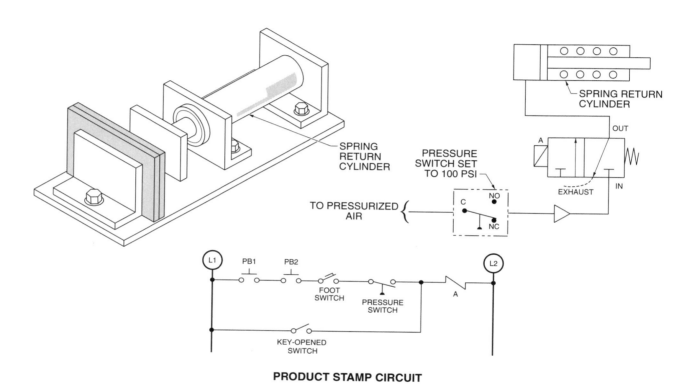

PRODUCT STAMP CIRCUIT

_____ 3. In the Conveyor Positioning Circuit, PB1 and PB2 are connected in ___ circuit logic with solenoid B.

_____ 4. In the Conveyor Positioning Circuit, PB1 and PB2 are connected in ___ circuit logic with solenoid A.

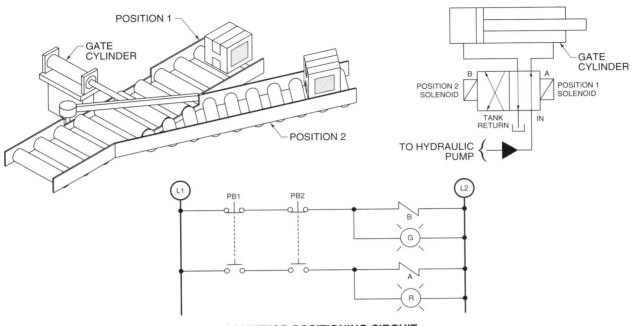

CONVEYOR POSITIONING CIRCUIT

_____ **5.** In the Glue Dispense Circuit, can the photoelectric and test button both energize the glue valve when the two-position selector switch is in either position?

_____ **6.** In the Glue Dispense Circuit, if a fuse were added in series with the glue valve solenoid to protect the circuit, what circuit logic would the fuse by itself add into the control circuit?

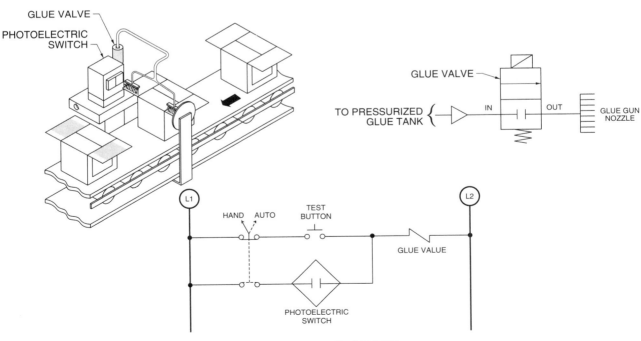

GLUE DISPENSE CIRCUIT

Activity 5-6: Troubleshooting Control Circuits

Answer the questions using the Motor Control Circuit.

_____ 1. The correct placement of DMM 1 so that the meter checks the fuse in Line 3 is position ___.

_____ 2. If the circuit is working correctly, can DMM 2 ever read 230 V at the motor when DMM 3 is reading 0 V?

_____ 3. If the circuit is working correctly, can DMM 3 ever read 230 V at the motor when DMM 2 is reading 0 V?

_____ 4. If DMM 4 reads 230 V in position 1 and 230 V in position 2, which fuse is bad?

_____ 5. If DMM 5 reads 230 V in position 1 and 0 V in position 2, which fuse is bad?

_____ 6. DMM 6 checks the fuse in Line ___.

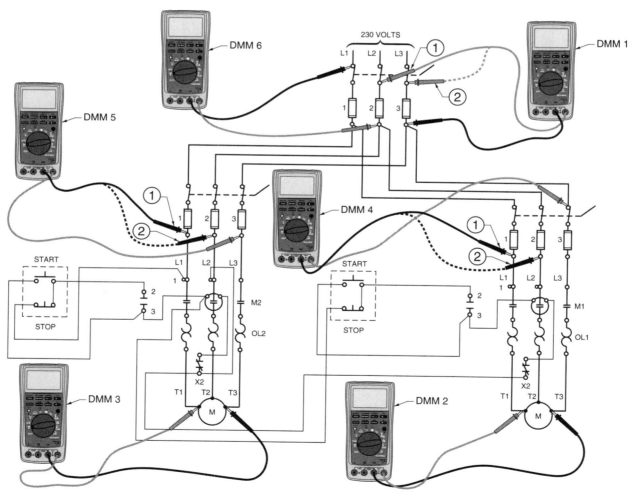

MOTOR CONTROL CIRCUIT

When troubleshooting control switches, approximate meter readings should be anticipated if the meter readings are going to be used to help determine circuit problems. Assume that the circuit is operating properly and determine the expected DMM readings.

_____ 7. The expected reading of DMM 1 with the pushbutton closed is ___ VAC.

_____ 8. The expected reading of DMM 1 with the pushbutton open is ___ VAC.

_____ 9. The expected reading of DMM 2 with the pushbutton closed is ___ VAC.

_____ 10. The expected reading of DMM 2 with the pushbutton open is ___ VAC.

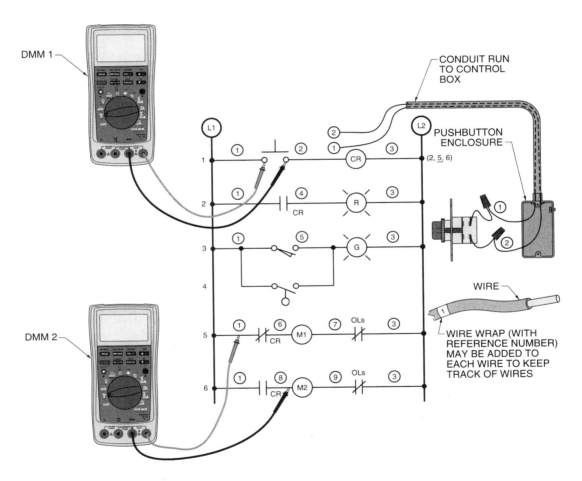

MOTOR CONTROL CIRCUIT

Assume that the circuit is operating properly and determine the expected DMM readings.

_____ **11.** The expected reading of DMM 1 when PL1 is ON and FL1 is closed is ___ VAC.

_____ **12.** The expected reading of DMM 1 when PL1 is ON and FL1 is open is ___ VAC.

_____ **13.** The expected reading of DMM 2 when PL1 is ON, PL2 is ON, and FL1 is closed is ___ VAC.

_____ **14.** The expected reading of DMM 2 when PL1 is ON, PL2 is OFF, and FL1 is closed is ___ VAC.

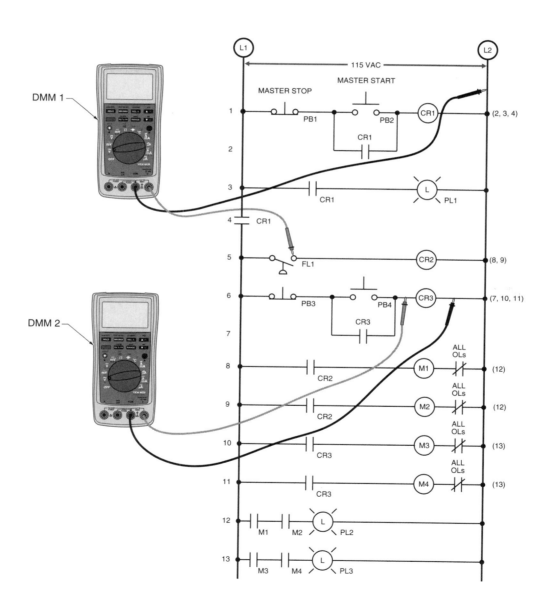

Connect each meter to the terminal strip numbered terminals to perform the required test.

15. Troubleshoot the transformer by connecting DMM 1 at the terminal strip to measure the control voltage out of the transformer and control circuit fuse. Troubleshoot the overload (OL) contact by connecting DMM 2 at the terminal strip to measure the voltage out of the fuse and grounded side of the transformer (to ensure that the control circuit is powered). Troubleshoot the overload contact by connecting DMM 3 to measure the voltage out of the fuse and out of the overload contact (to ensure that the overload contact is not tripped/opened).

_____ 16. The expected reading of DMM 1 if the circuit is working properly is ___ VAC.

_____ 17. The expected reading of DMM 2 if the circuit is working properly is ___ VAC.

_____ 18. The expected reading of DMM 3 if the circuit is working properly is ___ VAC.

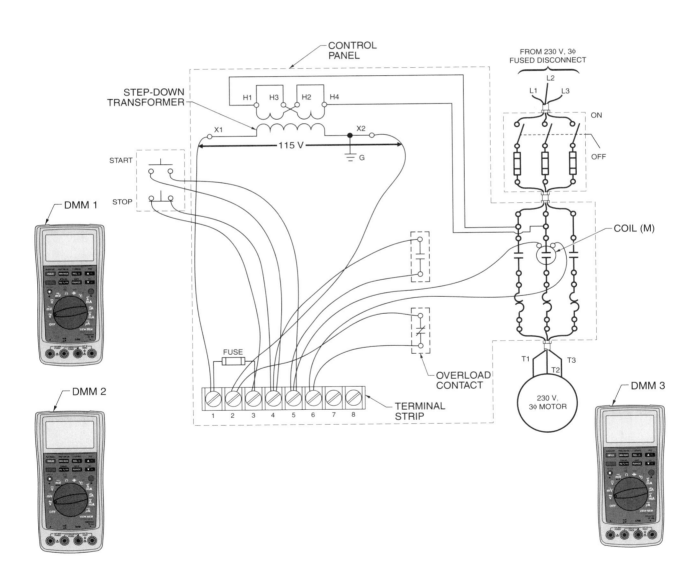

Answer the questions using the Control Transformer Circuit. The motor operates when the motor starter power contacts are manually closed by pressing them down at the motor starter. The motor does not operate when the start pushbutton is pressed or when a fused jumper test wire is connected from points 1 to 3. A DMM set to measure voltage connected at points X1 and X2 indicates proper voltage. A DMM set to measure voltage connected at points 3 and 6 indicates no voltage at any time.

_____ 19. The ___ is the most likely cause of the malfunction.

_____ 20. Is it likely the problem is in the overloads?

_____ 21. Is it likely the problem is in the motor starter coil?

_____ 22. Is it likely the problem is in the stop pushbutton?

_____ 23. Is it likely the problem is in the start pushbutton?

_____ 24. Is it likely the problem is in the transformer?

_____ 25. Is it likely the problem is in the control circuit fuse?

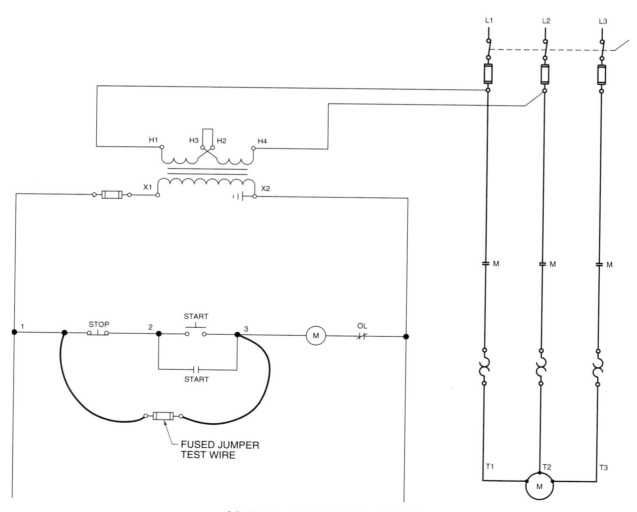

CONTROL TRANSFORMER CIRCUIT

Solenoids, DC Generators, and DC Motors

Applications 6

Application 6-1: Solenoids

Coils

A *solenoid* is an electric output device that converts electrical energy into a linear mechanical force. In a solenoid, an electric coil uses current to produce the power required to move a plunger. The movement of the plunger produces a linear mechanical force that is used to produce work in many industrial applications.

Coil Specifications

The current drawn by a solenoid coil depends on the applied voltage and size of the coil. Manufacturers list coil specifications to assist in installation and sizing of components. Because magnetic coils are encapsulated and cannot be repaired, they must be replaced when they fail. **See Coil Specifications.**

COIL SPECIFICATIONS

Size	Number of poles	Inrush current* 60 cycles					Sealed current* 60 cycles					Approximate operating time†	
		120 V	208 V	240 V	480 V	600 V	120 V	208 V	240 V	480 V	600 V	Pick-up	Drop-out
00	1-2-3	0.50	0.29	0.25	0.12	0.07	0.12	0.07	0.06	0.03	0.02	28	13
0	1-2-3-4	0.88	0.50	0.44	0.22	0.17	0.14	0.08	0.07	0.04	0.03	29	14
1	1-2-3-4	1.54	0.89	0.77	0.39	0.31	0.18	0.10	0.09	0.04	0.04	26	17
2	2-3-4	1.80	1.04	0.90	0.45	0.36	0.25	0.14	0.13	0.06	0.05	32	14
3	2-3	4.82	2.78	2.41	1.21	0.97	0.36	0.21	0.18	0.09	0.07	35	18
	4	5.34	3.08	2.67	1.33	1.07	0.39	0.23	0.20	0.10	0.08	35	18
4	2-3	8.30	4.80	4.15	2.08	1.66	0.54	0.31	0.27	0.14	0.11	41	18
	4	9.90	5.71	4.95	2.47	1.98	0.61	0.35	0.31	0.15	0.12	41	18
5	2-3	16.23	9.36	8.11	4.06	3.25	0.81	0.47	0.41	0.20	0.16	43	18

*in A
†in ms

For example, the coil used in a size 2 motor starter may have two, three, or four poles and draw 0.13 A sealed current when connected to a 240 V power source. This current value is used when selecting circuit fuse, wire, and transformer sizes. **See Motor Starter.**

The manufacturer-listed coil specifications are required when designing and troubleshooting a circuit. For example, when troubleshooting blown control circuit fuses, both the inrush and sealed current ratings of the coil must be considered. In a circuit that uses two magnetic motor starters (M1 and M2), it is possible for the two motor starters to be energized separately or at the same time. This situation (using two size 3 starters to control a three-phase motor with a 120 V control circuit) could cause the following current conditions:

- If M1 starts alone, there would be 4.82 A inrush current and 0.36 A sealed current.
- If M2 starts alone, there would be 4.82 A inrush current and 0.36 A sealed current.
- If M1 is ON and M2 is started, there would be 5.18 A as M2 starts and 0.72 A when both M1 and M2 are ON.
- If M2 is ON and M1 is started, there would be 5.18 A as M1 starts and 0.72 A when both M1 and M2 are ON.
- If M1 and M2 are started at the same time, there would be 10.36 A inrush current and 0.72 A sealed current.

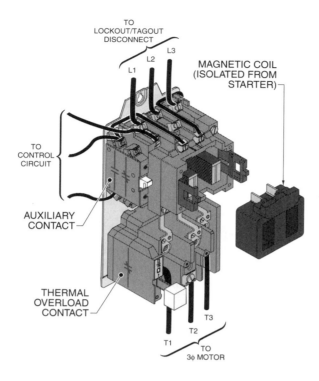

MOTOR STARTER

Solenoid-Operated Valves

Solenoid-operated valves are used to control the flow of fluids in electrically controlled systems. Solenoid-operated valves are used to control the direction of flow, the rate of flow, and the sequence of operation. **See Solenoid-Operated Valve.** When ordering and installing solenoid-operated valves, the requirements of the electrical part of the valve (coil) and the requirements of the fluid power part of the valve must be known. Additional information may be helpful in selecting the proper valve for other applications. Most manufacturers provide a specification order sheet that includes additional information to help in selecting the correct valve for an application.

Although the starting of both M1 and M2 simultaneously is unlikely, this situation must be considered as it could cause a 10 A (or less) fuse to blow. Understanding the circuit operation and the circuit component ratings is required during design and troubleshooting. **See Two Magnetic Motor Starters.**

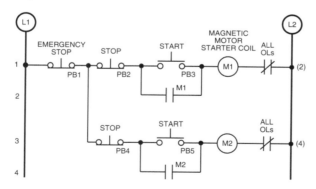

TWO MAGNETIC MOTOR STARTERS

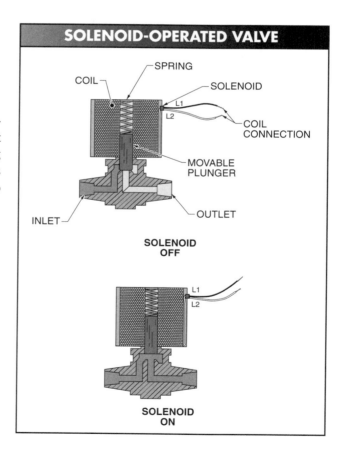

Application 6-2: Directional Control Valves

Directional Control Valves

Solenoid-operated valves typically control pneumatic and hydraulic equipment. A solenoid is used to move the valve spool that controls the flow of fluid in a directional control valve. A *directional control valve* is a valve used direct the flow of fluid in a fluid power system. Directional control valves do not affect the pressure or flow rate of the fluid. Directional control valves are identified by the number of positions, ways, and type of actuators.

Positions

Directional control valves are placed in different positions to start, stop, or change the direction of fluid flow. A *position* is the number of positions within the valve that the spool is placed in to direct fluid flow through the valve. Two- and three-way valves have two positions. Four-way valves may have two or three positions. **See Valve Positions.**

Ways

A *way* is the flow path through a valve. Most directional control valves are either two-way, three-way, or four-way valves. The number of ways required depends on the application. Two-way directional control valves have two main ports that allow or stop the flow of fluid. Two-way valves are used as shutoff, check, and quick exhaust valves. Three-way valves allow or stop fluid flow or exhaust. Three-way valves are used to control single-acting cylinders, fill and drain tanks, and control nonreversible fluid motors. Four-way directional control valves have four (or five) main ports that change fluid flow from one port to another. Four-way valves are used to control the direction of double-acting cylinders or reversible fluid motors. Important specifications to consider when choosing directional control valves include inlet and outlet diameter, working pressure, and operating temperature. Media types include gases, liquids, and liquids with suspended solids. **See Two-Way Valves.**

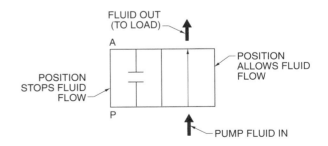

Actuators

Directional control valves must have a means to change the valve position. An *actuator* is a device that changes the valve position. Directional control valve actuators include pilots, solenoids, springs, manual levers, and palm buttons. Symbols are used to graphically describe the various fluid power valve actuators on system drawings so they can be identified quickly and accurately. **See Common Valve Actuators.**

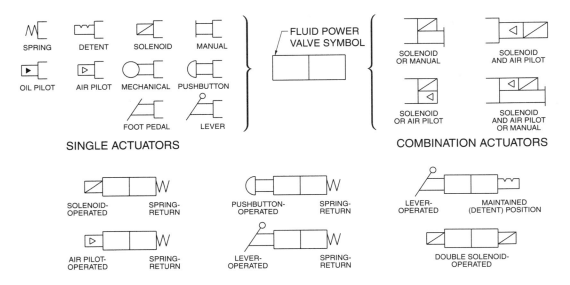

COMMON VALVE ACTUATORS

Valve Numbering System

Directional control valve manufacturers provide a numbering system to simplify valve selection and ordering. The numbering system consists of letters and numbers that represent the different valve models. A valve numbering system should include information on the number of ways, number of positions, port size, actuator and return methods, coil voltage, and current type. **See Valve Selections.**

For example, a Model 22AS2-120-1 valve is a four-way, two-position valve with 1/8" NPTF ports. The valve is solenoid-actuated, is solenoid-returned, and has a 120 VAC coil voltage.

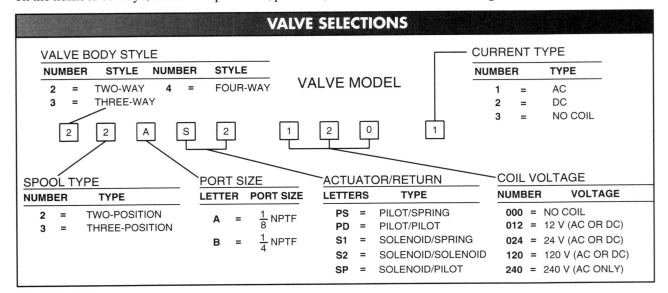

Application 6-3: Fluid Power Color Coding

Color Coding

The three basic types of power used to produce work are mechanical (shafts, gears, belts), electrical (motors, solenoids), and fluid power (hydraulic or pneumatic). Symbols are used for designing, operating, and maintaining fluid power systems. Fluid power symbols make the different components in the circuit easier to understand. Fluid power is used in industrial applications because it develops a large amount of power in a small space. Fluid power systems are typically found in all areas of industrial manufacturing and processing facilities. Fluid power circuits use color codes to identify operating conditions in a circuit. **See Selected Symbols** and **Fluid Power Color Code.**

SELECTED SYMBOLS

Cylinders	Hydraulic Pumps		Hydraulic Motors	
	Fixed-Displacement	Variable-Displacement	Fixed-Displacement	
DOUBLE-ACTING / DOUBLE-ACTING W/ DOUBLE END ROD	1 ROTATION	1 ROTATION	1 ROTATION	DUAL ROTATION

Directional Control Valves		Flow-Control Valves	Pressure-Control Valves
TWO-WAY NC	FOUR-WAY SOLENOID OPERATED		

Air or Oil Filter	Heat Exchanger	Pressure Gauge	Check Valve	Manual Shutoff

FLUID POWER COLOR CODE

Color	Meaning	Color	Meaning
Red	System operating pressure or highest working pressure. System pressure is the fluid flow after the pump until the flow is reduced, metered, or returned to the tank.	Yellow	Controlled flow by a metering device. Controlled flow is the fluid flow after a flow-control valve has reduced the volume (gpm) of fluid flow.
Blue	Exhaust or return flow back to the tank. Exhaust flow is the fluid flow from the actuator, back through the valve, and to the tank.	Orange	Reduced pressure that is lower than system operating pressure. Reduced flow is the fluid flow after a pressure-reducing valve has reduced the pressure (psi) of the fluid.
Green	Intake flow to pump or drain line flow. Intake flow is the fluid flow from the reservoir tank, through the filters, and to the pump.	Violet	Intensified pressure. Intensified pressure is pressure higher than system operating pressure.
		None	Inactive hydraulic fluid (reservoir fluid).

Application 6-4: Troubleshooting DC Motor Circuits

Troubleshooting DC Motor Circuits

When troubleshooting, approximate meter readings should be anticipated if the meter readings are going to have meaning and help determine circuit problems. The speed of a DC motor is determined by the voltage applied to the motor. A DC motor runs at full nameplate rated speed when the full nameplate rated voltage is applied to the motor. The less the applied voltage to the motor, the slower the motor speed. A starting rheostat is used to start the motor at reduced voltage by connecting resistance in series with the motor. The total applied voltage is divided between the rheostat and the motor. The higher the rheostat resistance, the greater the voltage drop across the resistor and thus less voltage is applied to the motor.

Solenoids, DC Generators, and DC Motors

6 Activities

Name _____ Date _____

Activity 6-1: Solenoids

Answer the questions using the Coil Specifications on page 69 for a 120 V, size 00, three-pole contactor.

_____ 1. The inrush current is ___ A.

_____ 2. The sealed current is ___ A.

_____ 3. The pick-up time is ___ ms.

_____ 4. The drop-out time is ___ ms.

Answer the questions using the Coil Specifications on page 69 for a 240 V, size 5, three-pole contactor.

_____ 5. The inrush current is ___ A.

_____ 6. The sealed current is ___ mA.

_____ 7. The pick-up time is ___ ms.

_____ 8. The drop-out time is ___ ms.

Answer the questions using the Coil Specifications on page 69 for a 120 V, size 3, three-pole contactor.

_____ 9. The inrush current is ___ A.

_____ 10. The sealed current is ___ mA.

_____ 11. The pick-up time is ___ ms.

_____ 12. The drop-out time is ___ ms.

Answer the questions using the Coil Specifications on page 69 for a 480 V, size 3, three-pole contactor.

_____ 13. The inrush current is ___ A.

_____ 14. The sealed current is ___ A.

_____ 15. The pick-up time is ___ ms.

_____ 16. The drop-out time is ___ ms.

Activity 6-2: Directional Control Valves

Complete the symbol for each of the stated valves.

1. Two-position, two-way, normally closed, solenoid-actuated, spring return valve

2. Two-position, two-way, normally open, solenoid-actuated, spring return valve

3. Two-position, two-way, double solenoid-actuated valve

4. Two-position, two-way, normally closed, manually or solenoid-actuated valve

5. Two-position, two-way, normally open, manually or solenoid-actuated valve

6. Two-position, two-way, double manually or double solenoid-actuated valve

List the model number using Valve Selections on page 72.

_____ 7. The model number for a three-way, two-position, ⅛″ NPTF, solenoid-actuated, spring-return, 120 VAC operated valve is ___.

_____ 8. The model number for a four-way, three-position, ⅛″ NPTF, solenoid-actuated, solenoid-return, 120 VDC operated valve is ___.

_____ 9. The model number for a two-way, two-position, ¼" NPTF, solenoid-actuated, pilot-return, 240 VAC operated valve is ___.

_____ 10. The model number for a three-way, two-position, ¼" NPTF, solenoid-actuated, spring-return, 24 VDC operated valve is ___.

_____ 11. The model number for a three-way, three-position, ¼" NPTF, solenoid-actuated, solenoid-return, 24 VAC operated valve is ___.

_____ 12. A model number ___ valve controls a double-acting cylinder. When one pushbutton is pressed and released, the cylinder advances, and when a second pushbutton is pressed and released, the cylinder retracts. The circuit is a 120 VAC and requires a ⅛" NPTF port.

Activity 6-3: Fluid Power Color Coding

Mark the color in each part of the fluid power circuit using the Fluid Power Color Code chart on page 73.

_____ 1. Color of A is ___.

_____ 2. Color of B is ___.

_____ 3. Color of C is ___.

_____ 4. Color of D is ___.

_____ 5. Color of E is ___.

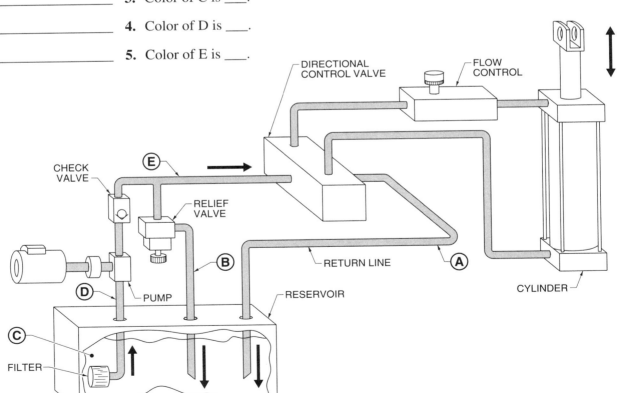

Identify the fluid power symbols on the Clamp and Eject Circuit.

_____ **6.** Two-way, cam-operated valve

_____ **7.** Pressure gauge

_____ **8.** Pressure-reducing valve

_____ **9.** Double-acting cylinder

_____ **10.** Needle valve

_____ **11.** Check valve

_____ **12.** Hydraulic pump

_____ **13.** Pressure-relief valve

_____ **14.** Filter

_____ **15.** Four-way, solenoid-operated valve

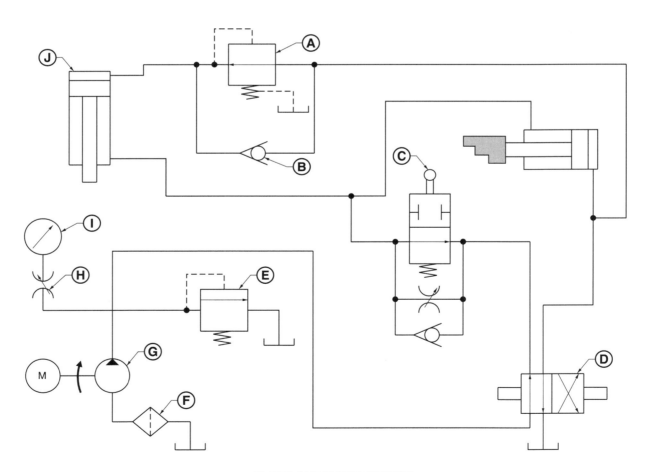

CLAMP AND EJECT CIRCUIT

Mark the color in each part of the fluid power circuit using the Fluid Power Color Code chart on page 73.

_____ **16.** Color of A is ___.

_____ **17.** Color of B is ___.

_____ **18.** Color of C is ___.

_____ **19.** Color of D is ___.

_____ **20.** Color of E is ___.

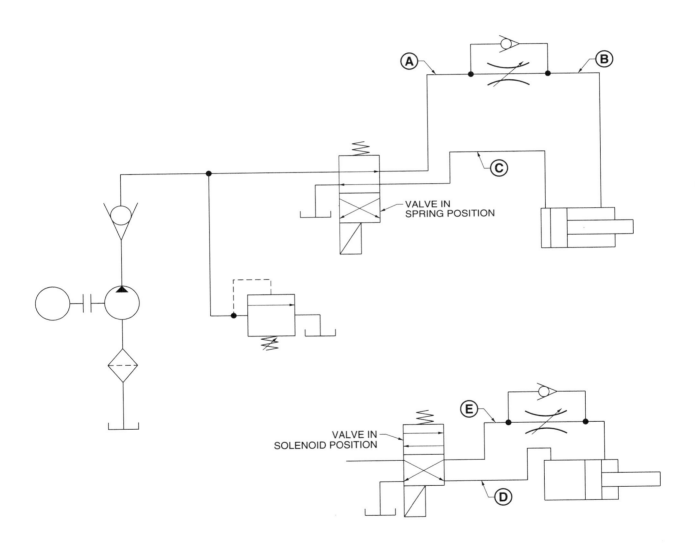

Activity 6-4: Troubleshooting DC Motor Circuits

Determine the expected DMM reading if the circuit is operating properly.

_____ 1. The expected reading of DMM 1 with the motor OFF is ___ VDC.

_____ 2. The expected reading of DMM 1 with the motor running at full speed is ___ VDC.

_____ 3. The expected reading of DMM 2 with the motor OFF is ___ VDC.

_____ 4. The expected reading of DMM 2 with the motor running at full speed is ___ VDC.

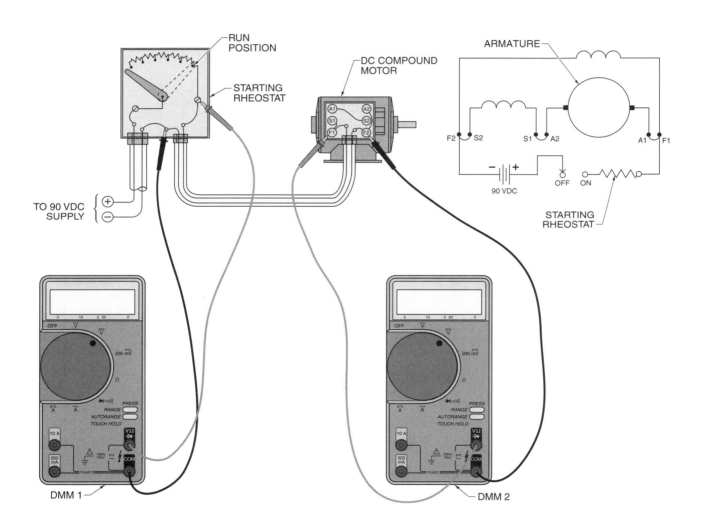

AC Generators, Transformers, and AC Motors

Applications

Application 7-1: AC Generators

Selecting Portable Generators

Portable generators are used to supply electrical power in emergencies, on construction sites, for recreation, and as a backup for utility power feeds. A properly sized portable generator should deliver enough power (in watts) at the correct voltage (12 VDC, 120 VAC, 120/240 VAC) to operate all loads connected to the generator. An undersized generator cannot operate the loads as required, which can damage the load and generator. An oversized generator supplies enough power to the loads, but is not energy efficient.

Portable generators are rated by their maximum power output, surge power output, and voltage output(s). **See Generators.** For example, a portable generator may be rated for 7500 W (7.5 kW) maximum power output and 10,750 W (10.75 kW) surge power output at 120 VAC. The surge power output rating is used to select a generator that has enough power to handle loads that include motors with a higher starting power than running power. For example, a 1 HP air compressor requires approximately 1500 W when running and approximately 5500 W when starting.

Determining Generator Size

To determine the correct generator for an application, apply the following procedure:

1. List the electrical devices that must be operated by the generator. This list should include devices likely to be operated (lamps, power tools, etc.) as well as devices that may be added (refrigeration units during prolonged power outages, sump pumps during heavy rains, etc.). The list of devices that must be operated by a generator determines the absolute minimum generator size that can be selected for an application. By also listing devices that may be added, the required generator size would be the most appropriate for the application.

Baldor Electric Co.

		GENERATORS	
Model	Maximum Power*	Surge Power*	Voltage Output
1-A	3500	7500	120 VAC GFCI Duplex Outlet
2-A	4500	9000	120 VAC GFCI Duplex Outlet 120/240 VAC Twistlock Outlet
3-A	7500	10,750	12 VDC (2) 120 VAC GFCI Duplex Outlets 120/240 VAC Twistlock Outlet
4-A	10,000	17,000	12 VDC (3) 120 VAC GFCI Duplex Outlets 120/240 VAC Twistlock Outlet
5-A	12,500	21,000	12 VDC (4) 120 VAC GFCI Duplex Outlets 120/240 VAC Twistlock Outlet

* in W

2. Determine the voltage requirements of the loads connected to the generator. The voltage output of a generator must be within +5%/–10% of the load(s) connected to the generator. Lamps, portable power tools, and most small appliances are rated at 115/120 VAC. Most large power-consuming devices (electric stoves/heaters, air compressors, etc.) are rated at 230/240 VAC. If a device is rated for only one voltage, it cannot be connected to any other voltage. For example, a motor rated for 230 VAC must be connected to a 230 VAC (+5%/–10%) power supply.

However, a motor rated for 115/230 VAC may be connected to either a 115 VAC or 230 VAC power supply.

A 115/230 VAC motor (or other load) draws the same amount of power when connected to 115 VAC or 230 VAC. However, the current draw at the higher voltage (230 VAC) is half that of the current draw at the lower voltage (115 VAC) because power (P) is equal to voltage (E) times current (I). Thus, wiring a dual-voltage load at a higher voltage reduces the required wire size and increases the permissible wire length, but does not reduce (or increase) the generator required power output.

3. Determine the operating power requirements of each load and the total power requirements of all the loads that are connected to the generator. Some devices, such as lamps and heating elements, have a power listing in watts (100 W, 500 W, etc.). Other loads such as motors, some tools, and appliances do not have a power rating listed but have a current rating listed. The power rating of a load that has a listed voltage and current rating can be determined by multiplying the rated voltage by the rated current.

4. Identify all the loads to be connected to the generator that have a starting power draw higher than the operating power. This usually includes all loads that contain a motor. In general, any load that includes a motor that is connected directly to a load can be calculated as having a high starting power requirement. This includes motors that are directly coupled through belts, chains, and gears to loads such as fans, pumps, and tools. However, small motors such as those used in portable cooling fans and cooling motors in computers, etc., need not be considered as having a high starting power draw.

Problems occur if the high starting power requirements of loads are not factored into the generator size requirements. The problem occurs when a large amount of starting power is drawn from the generator, as with air compressor starting, etc. In this case, the total voltage output of the generator drops if the generator is fully loaded. The larger the overload, the larger the voltage drop. Even a temporary low voltage on computers, appliances, tools, etc., can cause damage to the loads and generator.

The wattage rating listed on a load nameplate can be used to determine the total wattage requirements of a generator. When load nameplate ratings are not available, average wattage requirement guides can be used to determine total wattage requirements. **See Average Wattage Requirement Guide. See Appendix.**

AVERAGE WATTAGE REQUIREMENT GUIDE

Electrical Load	Operating Power*	Starting Power*
Construction site		
Air Compressor (½ HP)	1500	5500
Air Compressor (1 HP)	3000	11,000
Electric Welder (200 A)	9000	9000
High-Pressure Washer (1 HP)	1200	3600
Circular Saw (7¼")	1400	2300
Table Saw (10")	1800	4500
Hand Drill (½")	600	800
Grinder (4½")	750	950
Grinder (6")	1000	1300
Grinder (9")	2300	3000
Hand Jigsaw	650	850
Reciprocating Saw (7" blade)	1150	1600
Sander (⅓ sandpaper sheet size)	350	550
Sander (½ sandpaper sheet size)	450	650
Battery Charger (15 A, no boost)	375	375
Battery Charger (60 A, no boost)	1500	1500
Battery Charger (100 A, no boost)	2500	2500

* in W

5. Once the total power requirements are determined, a generator can be selected. The minimum generator size should be based on the maximum possible power draw for the given application and the amount of extra (spare) power desired. Typically, 25% to 50% additional power is recommended as a minimum amount of extra power. This ensures that a generator does not operate at 100% capacity, which shortens generator life and/or lowers the generator voltage output, and there is extra power for the times that may require additional power.

If the total power required is high for any one application, such as a large construction site, etc., it is better to use more than one generator and divide the loads between the generators. This also ensures that there is available power if one of the generators has a problem.

Application 7-2: Transformers

Dual-Voltage Control Transformer Operation

In most industrial applications, the motor power circuit is at a high voltage that produces the required power at the motor. The control circuit is at a low voltage because it is safer than high voltage. A control transformer is used to reduce the high voltage of the power circuit to a low voltage that can be used for the control circuit.

Most control transformers have a dual-voltage primary. A dual-voltage primary allows the transformer to be connected to a 240 V or 480 V power circuit. The primary coils of the transformer are connected in series for high voltage and in parallel for low voltage. The secondary side of the transformer delivers a low-voltage output (normally 120 V) for the control circuit. **See Dual-Voltage Transformer** and **Dual-Voltage Transformer Application.**

Transformer Tap Connections

A transformer with taps is used to compensate for voltage differences in a control circuit. Taps are connecting points that are provided along a transformer coil. Taps are usually provided at 2½° increments along one end of the transformer coil. **See Transformer Nameplate.**

To determine the proper tap connections, apply the following procedure:

1. Measure the incoming voltage on the primary side of the transformer using a DMM set to measure voltage.
2. Determine the secondary voltage. The secondary voltage is the voltage rating of the load(s) that are connected to the transformer.
3. Determine the connections using the transformer nameplate or specifications.
4. Shut the power OFF, lock out, and tag the incoming power to the transformer. Ensure the power is OFF by testing the circuit with a DMM set to measure voltage.
5. Connect the transformer as listed by the manufacturer. Check each connection twice.
6. Turn the power ON.
7. Measure the secondary voltage of the transformer. If the secondary voltage is not correct, repeat Steps 1 – 6.

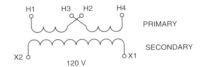

CONNECTIONS FOR DUAL-VOLTAGE TRANSFORMER

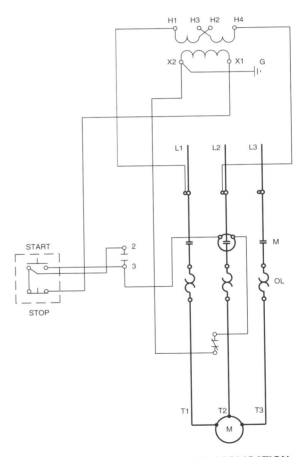

DUAL-VOLTAGE TRANSFORMER

DUAL-VOLTAGE TRANSFORMER APPLICATION

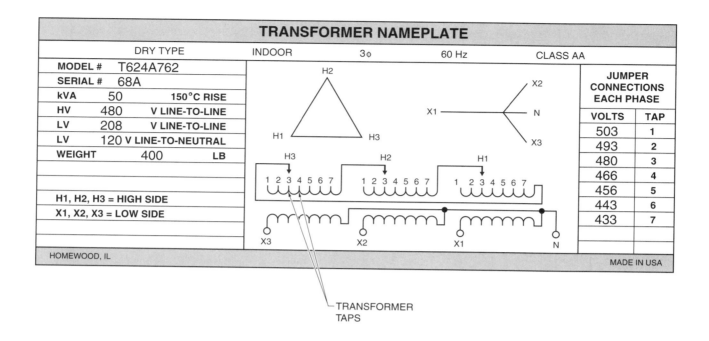

TRANSFORMER TAPS

Application 7-3: Troubleshooting AC Motors and Motor Circuits

Troubleshooting Control Transformer Section

When troubleshooting, approximate meter readings should be anticipated before a meter is connected into a circuit if the meter readings are going to have meaning and help determine circuit problems. When troubleshooting an AC motor circuit, the troubleshooting process is started by taking voltage measurements at the control transformer. The voltage at the control transformer primary can be measured to verify there is power in the high-voltage power circuit. If there is no voltage present, the fuses/circuit breakers in the main disconnect are checked. If the voltage reading is correct, the voltage out of the transformer (secondary side) is checked. If the voltage reading is not correct, the fuses/circuit breaker on the transformer are checked.

Troubleshooting Split-Phase Motors

When troubleshooting a motor, resistance measurements are taken with all power OFF. In a split-phase motor, resistance measurements can be taken at the starting winding and running winding. The starting winding has a higher resistance than the running winding because the starting winding is usually made of smaller gauge wire than the running winding. The smaller gauge wire is used because the winding is only energized during starting.

Troubleshooting Main Disconnect Fuses

Fuses and circuit breakers are used to create an intentionally weak point in a circuit that opens when an overcurrent condition occurs. For this reason, the fuses/circuit breakers should be checked when troubleshooting a circuit. Each fuse must be checked because it is possible to have one or more fuses open (blown) without having all the fuses in a disconnect (or circuit) open.

AC Generators, Transformers, and AC Motors

Activities 7

Name _____ Date _____

Activity 7-1: AC Generators

Answer the questions using Generators on page 81 and Average Wattage Requirement Guide in Appendix. A construction site requires one 120/230 VAC, ½ HP air compressor; six 120 VAC, 7¼" circular saws; one 240 VAC, 10" table saw; two 120 VAC, ½" hand drills; four 100 W lamps; one 120 VAC, ½ HP sump pump; and one 50 W radio.

_____ 1. Is a 120 VAC, 230 VAC, or 120/230 VAC generator required for the construction site loads?

_____ 2. The total maximum operating power draw at any one time is ___ W.

_____ 3. The total starting power draw if all loads are started at the same time is ___ W.

Determine the total starting power draw if:
- *Only two of the six circular saws are typically started at any one time, but all six can be operating at one time*
- *Only one hand drill is likely to be started at any one time*
- *The air compressor, table saw, and sump pump cycle ON and OFF, and thus must be included in the starting power draw calculation. However, since it is unlikely that all three will start at precisely the same time, use the starting power draw of the two higher power devices and the operating power of the lowest power device to determine starting power draw.*

_____ 4. Total starting power draw is ___ W.

Determine if each portable generator combination can supply enough power to operate all the loads on the construction site and provide an additional 25% reserve power above the maximum calculated value.

_____ 5. One model 5-A portable generator

_____ 6. Two model 3-A portable generators

_____ 7. Two model 4-A portable generators

_____ 8. One model 5-A portable generator and one model 4-A portable generator

_____ 9. Three model 3-A portable generators

_____ 10. Five model 1-A portable generators

Answer the questions using Generators on page 81 and Average Wattage Requirement Guide in Appendix. A portable generator must be selected for a residence to protect against a possible total power utility outage. Assume that each load to be powered by the portable generator can be easily (and safely) switched from the utility power circuit to the portable generator. The minimum portable generator power requirements to maintain power during a power outage are 600 W of lighting; one 120 VAC, ⅓ HP sump pump; one 19" television; one standard microwave oven; one electric skillet; two 120 VAC, 1250 W space heaters; and one home security system.

_____ 11. Is a 120 VAC, 230 VAC, or 120/230 VAC generator required for the residential loads?

_____ 12. The total maximum operating power draw at any one time is ___ W.

_____ 13. The total starting power draw if all loads are started at the same time is ___ W.

Determine if each portable generator can supply enough power to operate all the loads during the power outage and provide an additional 50% reserve power above the maximum calculated value (so additional loads can be added).

_____ 14. One model 1-A portable generator

_____ 15. One model 2-A portable generator

_____ 16. One model 3-A portable generator

_____ 17. One model 4-A portable generator

_____ 18. One model 5-A portable generator

Activity 7-2: Transformers

Connect the transformer taps for each primary voltage requirement using Primary Tap Connections. Connect the transformer taps for each secondary voltage requirement using Secondary Tap Connections.

PRIMARY TAP CONNECTIONS		
Primary volts	Connect primary lines to	Interconnect
190	H1 & H7	H1 to H6 H2 to H7
200	H1 & H8	H1 to H6 H3 to H8
208	H1 & H9	H1 to H6 H4 to H9
220	H1 & H10	H1 to H6 H5 to H10
380	H1 & H7	H2 & H6
400	H1 & H8	H3 & H6
416	H1 & H9	H4 & H6
440	H1 & H10	H5 & H6

SECONDARY TAP CONNECTIONS		
Secondary volts	Connect secondary lines to	Interconnect
220	X1-X4	X2 to X3
110/220	X1-X2-X4	X2 to X3
110	X1-X4	X1 to X3 X2 to X4

1. A transformer has a primary voltage of 416 V. The required secondary voltage is 220 V.

2. A transformer has a primary voltage of 208 V. The required secondary voltage is 110 V.

3. A transformer has a primary voltage of 190 V. The required secondary voltage is 110 V.

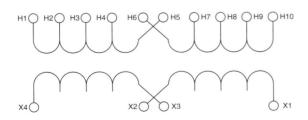

4. A transformer has a primary voltage of 440 V. The required secondary voltage is 220 V.

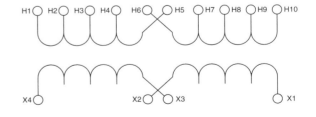

Activity 7-3: Troubleshooting AC Motors and Motor Circuits

Determine the expected DMM readings if the circuit is working properly.

_____ 1. The expected reading of DMM 1 when the motor is ON is ___ VAC.

_____ 2. The expected reading of DMM 1 when the motor is OFF is ___ VAC.

_____ 3. The expected reading of DMM 2 when the motor is ON is ___ VAC.

_____ 4. The expected reading of DMM 2 when the motor is OFF is ___ VAC.

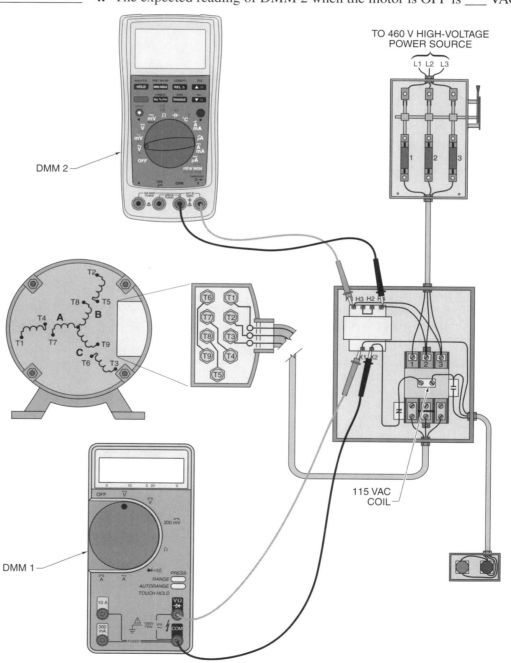

5. Draw the correct position of the selector switch on DMM 1 and DMM 2 to measure the resistance of the motor windings. Connect DMM 1 to measure the resistance of the running winding. Connect DMM 2 to measure the resistance of the starting winding.

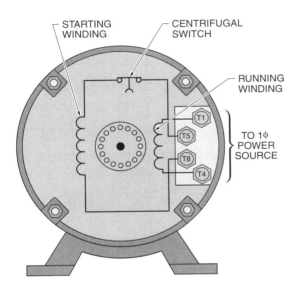

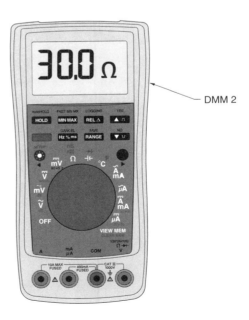

Determine the expected DMM readings if the circuit is working properly.

_____ **6.** The expected reading of DMM 1 when the motor is ON is ___ VAC.

_____ **7.** The expected reading of DMM 1 when the motor is OFF is ___ VAC.

_____ **8.** The expected reading of DMM 2 when the motor is ON is ___ VAC.

_____ **9.** The expected reading of DMM 2 when the motor is OFF is ___ VAC.

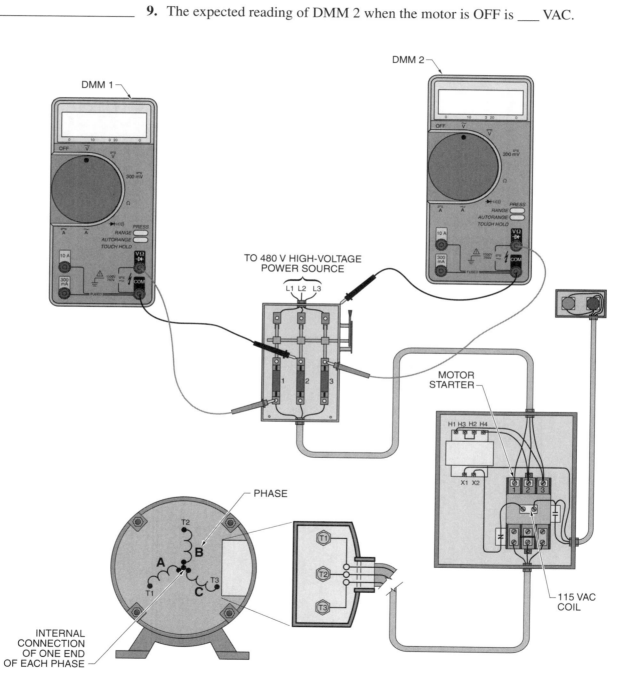

Contactors and Motor Starters

Applications 8

Application 8-1: Ambient Temperature Compensation with Overloads

Thermal Overloads

Thermal overloads are heat-sensing devices that provide a means of monitoring the current drawn by a motor. The thermal overloads trip when the heat generated by motor windings approaches a damaging level. In applications where the ambient temperature varies above or below the standard rating temperature of 104°F, ambient temperature must be accounted for because thermal overloads are temperature dependent. If the ambient temperature of a motor is different than that of the overloads, the overloads can cause nuisance tripping or motor burnout.

For example, if nontemperature-compensated overloads are used for a well pump motor that is at a different temperature than the surface-mounted overloads, the overloads must be adjusted for the difference in temperature. Graphs show temperature adjustments for different ambient temperatures. **See Heater Ambient Temperature Correction.**

To find the overload trip current using an ambient temperature correction chart, apply the following procedure:

1. Determine ambient temperature.
2. Find the rated current (%) for the ambient temperature on the graph.
3. Multiply the motor full-load current (from motor nameplate) by the rated current.

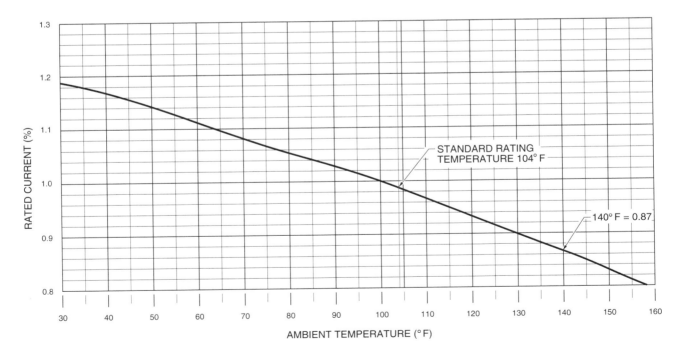

HEATER AMBIENT TEMPERATURE CORRECTION

Example: Finding Overload Trip Current

What is the overload trip current of a motor with a full-load current of 25 A that operates in an ambient temperature of 130°F?

1. Determine ambient temperature.

   Ambient temperature is 130°F.

2. Find the rated current (%) for the ambient temperature on the graph.

   At 130°F the rated current is 0.9 (from graph).

3. Multiply the motor full-load current (from motor nameplate) by the rated current.

   Overload trip current = 25 × 0.9
   Overload trip current = **22.5 A**

Application 8-2: Overload Trip Time

Overload Relays

Overload relays (OLs) are designed to protect a motor when it is running by tripping when an overload condition occurs, which disconnects the motor from the power lines. Overload relays allow temporary overloads (current higher than overload setting) for short periods of time to allow motors to start. This is required because motors draw a high starting current. Overload relays also trip and turn the motor OFF (by turning off the motor starter) if an overload exists for a long enough period of time to cause damage to the motor. All overload relays include a normally closed contact and many also include a normally open contact. The normally closed contact is used to turn OFF the starter, and the normally open contact can be used to sound an alarm (or turn ON a lamp) once the overload trips. Overload relays can be reset once the overload is removed.

The overload trip time depends on the extent of the overload. High overloads induce fast trip times. Manufacturers provide heater overload trip characteristic charts. **See Heater Trip Characteristics in Appendix.** For example, if a motor is overloaded 300%, the overloads trip in 58 sec (from Heater Trip Characteristics chart).

Overload Class Ratings

Overload relays are rated according to their trip class (Class 10, 20, 30, etc.). The class rating defines the length of time it takes for the overload relay to trip. The motor application must be considered when selecting an overload relay class. If a fast trip time is required, a low class number overload relay should be selected. **See Overload Class Trip Chart.**

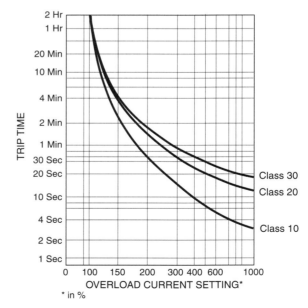

OVERLOAD CLASS TRIP CHART

Application 8-3: Overload Heater Size

Overload Heaters

The current rating on a motor nameplate is used to select overload heaters. If the exact current rating is not known, the motor full-load current is used to find the current rating. **See Full-load Currents in Appendix.**

Manufacturers provide overload heater selection charts with motor starters. The heater number is found by matching the motor nameplate-rated current value to the size of the starter. **See Heater Selections in Appendix.** For example, if a size 1 starter is used to control a motor with a 9 A rating, a #51 overload heater is required (from Heater Selections chart).

Application 8-4: Contactor and Motor Starter Ratings

Contactors and Motor Starters

Contactors and motor starters (contactors with added overloads) are rated according to the power and current they switch. Manufacturers provide ratings of different size motor starters. **See Control Ratings in Appendix.**

NEMA and IEC Ratings

Contactors and motor starters are rated according to the size (horsepower and/or current rating) and their voltage rating. The National Electrical Manufacturers Association (NEMA) and the International Electrotechnical Commission (IEC) are two primary organizations that rate contactors and motor starters.

NEMA contactors and motor starters are based on their continuous current (amperage) rating and voltage rating. NEMA ratings are listed as a size number, which ranges from size 00 to size 9. **See NEMA Contactor and Starter Ratings.**

IEC contactors and motor starters are based on their maximum operational current (amperage) rating. The IEC does not specify a size number. Instead, IEC contactors and motor starters state a utilization category rating that defines the typical duty of the IEC contactor or motor starter. Utilization categories AC-3 and AC-4 are used for most motor applications. **See IEC Contactor and Starter Ratings.**

In addition to the standard NEMA ratings, some motor starter manufacturers also provide motor matched

NEMA CONTACTOR AND STARTER RATINGS

NEMA Size	Continuous Current Rating*	Horsepower 230 VAC	Horsepower 460 VAC
00	9	1	2
0	18	3	5
1	27	7	10
2	45	15	25
3	90	30	50
4	135	50	100
5	270	100	200
6	540	200	400
7	810	300	600
8	1215	450	900
9	2250	800	1600

* in A

IEC CONTACTOR AND STARTER RATINGS

Utilization Category	IEC Category Description
AC-1	Used with noninductive or slightly inductive loads such as lamps and heating elements.
AC-2	Used with light inductive loads such as solenoids.
AC-3	Used with motors in which the motor is typically turned OFF only after the motor is operating at full speed.
AC-4	Used with motors in which the motor is used with rapid starting and stopping (jogging, inching, plugging, etc.).

sizes (MM sizes). Motor matched sizes fall between the standard NEMA sizes. Motor matched sizes allow for a more closely matched size in applications in which the motor size is known and some cost savings can be gained by using a half-size rated starter. **See Motor Matched Size Ratings.**

MOTOR MATCHED SIZE RATINGS

MM Size	Continuous Current Rating*	Horsepower 230 VAC	Horsepower 460 VAC
1¾	40	10	15
2½	60	20	31
3½	115	40	75
4½	210	75	150

* in A

Application 8-5: Sizing Motor Protection, Motor Starters, and Wire

Motor Protection

A high inrush current occurs when a motor is energized. This inrush current is many times the normal running current of the motor. Motors require special overload protection devices that can withstand the high starting current and still protect the motor from a sustained overload. Dual-element fuses are used to protect motors from overcurrent and short circuits.

Manufacturers provide charts for selecting the correct size and type of overload protection for a motor. These charts include information such as sizing backup protection, switch size, starter size, temperature rating, and wire and conduit sizes.

Single-Phase Motor Charts

A single-phase motor chart is used with 115 V and 230 V, 1ϕ motors up to 10 HP. **See 1ϕ Motors and Circuits. See Appendix.** Column 1 lists the horsepower and ampere rating for motors operating at normal speeds. Column 2 lists the fuse size for different motor temperature ratings and service factors. Column 3 lists the switch or fuse holder size. The size listed denotes the minimum ampere rating. Column 4 lists the minimum motor starter required when controlling the motor. The motor starter usually provides overload protection in addition to controlling the motor. Columns 5 and 6 list the controller termination temperature rating and minimum size copper wire and conduit required when connecting the motor. A bullet in Column 5 denotes insulation available. Column 6 is used in conjunction with Column 5 because wire type (TW, THW, etc.) and wire size must be considered.

Three-Phase Motor Charts

A three-phase motor chart is used with 230 V and 460 V, 3ϕ motors up to 300 HP. Once the motor size is determined, the chart is used in the same manner as the single-phase motor chart. **See 3ϕ Motors and Circuits in Appendix.**

SINGLE-PHASE MOTORS AND CIRCUITS

1		2		3	4	5				6	
Size of motor Table 430.248		Motor overload protection low-peak or Fusetron®				Controller termination temperature rating				Minimum size of copper wire and trade conduit	
						60°C		75°C			
HP 115 V	Amp (120 V system)	Motor less than 40°C or greater than 1.15 SF (Max fuse 125%)	All other motors (Max fuse 115%)	Switch 115% minimum or HP rated or fuse holder size	Minimum size of starter	TW	THW	TW	THW	Wire size (AWG or kcmil)	Conduit (in.)
⅙	4.4	5	5	30	00	•	•	•	•	14	½
¼	5.8	7	6½	30	00	•	•	•	•	14	½
⅓	7.2	9	8	30	00	•	•	•	•	14	½
½	9.8	12	10	30	00	•	•	•	•	14	½
¾	13.8	15	15	30	00	•	•	•	•	14	½
1	16.0	20	17½	30	00	•	•	•	•	14	½

Direct Current Motor Charts

A direct current (DC) motor chart is used with 90 V, 120 V, and 180 V motors up to 10 HP. Once the motor size is determined, the chart is used in the same manner as the single-phase motor chart. **See DC Motors and Circuits in Appendix.**

Application 8-6: Motor Starter Replacement Parts

Manufacturer Service Bulletins

A manufacturer service bulletin is used to order motor starter replacement parts. A *service bulletin* is a drawing and parts list for a device. The service bulletin contains the part numbers of components that may require replacement. Service bulletins should be kept for each piece of equipment and are included when the equipment is purchased. If a service bulletin is not included, one may be obtained from the manufacturer. **See Service Bulletin.**

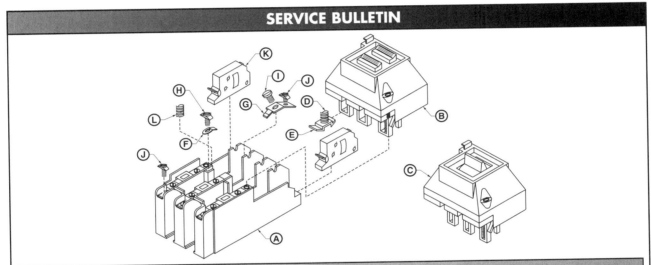

SERVICE BULLETIN

PARTS LIST

Item	Description	Part number	Size M-0 2-pole	Size M-0 3-pole	Size M-1 2-pole	Size M-1 3-pole	Size M-1P 2-pole
A	Contact Block	00401	1	1	1	1	1
B	Contact Actuator—Square pushbutton						
	2-pole	01753	1	—	1	—	1
	2-pole with run-jog feature	01752	1	—	1	—	—
	3-pole	01751	—	1	—	1	—
	3-pole with run-jog feature	01750	—	1	—	1	—
C	Contact actuator—Toggle						
	2-pole	35953	1	—	1	—	1
	2-pole with run-jog feature	35952	1	—	1	—	1
	3-pole	35951	—	1	—	1	—
	3-pole with run-jog feature	35950	—	1	—	1	—
D	Contact spring	10902	1	1	—	—	—
E	Movable contact	1040	1	1	—	—	—
F	Stationary contact—Load side	1095	—	—	1	1	1
G	Stationary contact—Line side	17310	—	—	1	1	1
H	Screw assembly #6-32 × ⅜″	12080	2	3	2	3	2
I	Screw assembly #8-32 × ⅜″	14121	2	3	2	3	2
J	Wire clamp screw assembly	01850	4	6	4	6	4
K	Internal interlock (NO)	2065	—	—	—	—	—
	Internal interlock (NC)	2066	—	—	—	—	—
L	Return spring	17901	2	2	2	2	2

Application 8-7: Motor Drives

Controlling Three-Phase, Low-Voltage, Delta-Connected Motors

When connecting a motor to the power source using a magnetic motor starter, the motor is connected to the full supply voltage at all times (starting and running). When connecting a motor to the power source using a motor drive, the motor can be connected at full supply voltage or a voltage less than full voltage (soft start). In addition to being able to control the voltage at the motor, a motor drive also controls the speed of the motor.

When using a magnetic motor starter to control a motor, a control circuit powered by a control transformer is generally used to energize the magnetic motor starter coil. When using a motor drive to control a motor, the control components (pushbuttons, flow switches, temperature switches, etc.) are connected directly to the motor drive control terminals.

Controlling Three-Phase, High-Voltage, Delta-Connected Motors

When connecting a motor to either a magnetic motor starter or motor drive, the motor must be connected for the applied voltage. If the delivered voltage of the starter or drive is low voltage (208 V–240 V), the motor is connected to support low voltage. If the delivered voltage of the starter or drive is high voltage (460 V–480 V), the motor is connected to support high voltage. Regardless of which voltage a motor is connected for (high or low), the control circuit (at drive control terminals) and the control components (pushbuttons, pressure switches, etc.) remain wired the same. A wiring diagram is used to show the terminal numbering system for a high-voltage, delta-connected, 3ϕ motor. The leads are marked T1 through T9 and a terminal connection chart is provided for wiring high- and low-voltage operation. The nine leads are connected in either series or parallel for high or low voltage.

Application 8-8: Troubleshooting Contactors

Troubleshooting Heating Circuits Using Wiring Diagrams

A contactor is the same as a motor starter without overloads (heaters). A contactor is used to control lamp loads and heating elements. As with motor control circuits, heating circuits must also include a control circuit. A step-down control transformer is used to lower the power circuit voltage to the level required in the control circuit.

An ON-delay (operate-delay relay) may be used to delay the turning on of a heating element after the temperature switch calls for heat. An OFF-delay (release-delay relay) may be used to prevent the heating element from turning back on for a set time period after it has shut off. This prevents the heating element from rapidly cycling on and off.

If a line diagram is not available when troubleshooting a circuit, the control circuit wiring diagram may be used to determine meter connection locations. In a wiring diagram, the location of components is generally shown to be as close as possible to the actual circuit configuration. If troubleshooting using a wiring diagram, it is recommended that the wiring diagram be converted into a line diagram to aid in locating test points in the circuit.

Troubleshooting Heating Circuits Using Line and Wiring Diagrams

When troubleshooting a heating element circuit, measurements must be taken in both the power circuit and the control circuit by using the same method used for troubleshooting a motor control circuit. The power circuit delivers the high voltage and current to the heating elements to produce the required power output. The control circuit operation determines when the heating elements are ON and OFF. A control transformer is used to reduce the voltage of the power circuit to the level required in the control circuit.

Contactors and Motor Starters

Activities 8

Name _____ Date _____

Activity 8-1: Ambient Temperature Compensation with Overloads

Determine the overload trip current for each motor installation using the Heater Ambient Temperature Correction on page 91. Note: Each motor is installed in a standard enclosure.

_____ 1. A 25 A rated motor is installed in a 104°F ambient temperature. The overload trip current is ___ A.

_____ 2. A 35 A rated motor is installed in a 50°F ambient temperature. The overload trip current is ___ A.

_____ 3. A 50 A rated motor is installed in a 140°F ambient temperature. The overload trip current is ___ A.

_____ 4. A 75 A rated motor is installed in a 77°F ambient temperature. The overload trip current is ___ A.

_____ 5. A 12.5 A rated motor is installed in a 149°F ambient temperature. The overload trip current is ___ A.

Activity 8-2: Overload Trip Time

Determine whether the overloads will or will not trip in the given situations. If they will trip, determine the overload trip time using the Heater Trip Characteristics in Appendix.

_____ 1. If a motor draws 100% of rated current, the overloads ___ trip in ___ sec.

_____ 2. If a motor draws 500% of rated current, the overloads ___ trip in ___ sec.

_____ 3. If a motor draws 50% of rated current, the overloads ___ trip in ___ sec.

_____ 4. If a motor draws 1000% of rated current, the overloads ___ trip in ___ sec.

_____ 5. If a motor draws 200% of rated current, the overloads ___ trip in ___ sec.

Determine whether the overloads will or will not trip in the given situations. If they will trip, determine the overload trip time using the Overload Class Trip Chart on page 92.

_____ 6. If the current draw by a motor is 200% of the overload current setting, a Class 10 overload ___ trip in ___ sec.

_____ 7. If the current draw by a motor is 200% of the overload current setting, a Class 20 overload ___ trip in ___ sec.

_____ 8. If the current draw by a motor is 200% of the overload current setting, a Class 30 overload ___ trip in ___ sec.

_____ 9. If the current draw by a motor is 75% of the overload current setting, a Class 10 overload ___ trip in ___ sec.

_____ 10. If the current draw by a motor is 400% of the overload current setting, a Class 30 overload ___ trip in ___ sec.

_____ 11. If the current draw by a motor is 700% of the overload current setting, a Class 10 overload ___ trip in ___ sec.

_____ 12. If the current draw by a motor is 150% of the overload current setting, a Class 10 overload ___ trip in ___ sec.

_____ 13. If the current draw by a motor is 150% of the overload current setting, a Class 30 overload ___ trip in ___ sec.

Activity 8-3: Overload Heater Size

Determine the heater size using Full-load Currents and Heater Selections in Appendix.

_____ 1. A number ___ heater is used with a ¾ HP, 115 V, 1φ motor with a size 0 starter.

_____ 2. A number ___ heater is used with a ¾ HP, 230 V, 1φ motor with a size 0 starter.

_____ 3. A number ___ heater is used with a 10 HP, 115 V, 1φ motor with a size 4 starter.

_____ 4. A number ___ heater is used with a 2 HP, 120 VDC motor with a size 0 starter.

_____ 5. A number ___ heater is used with a 2 HP, 120 VDC motor with a size 1 starter.

_____ 6. A number ___ heater is used with a 2 HP, 120 VDC motor with a size 2 starter.

_____ 7. A number ___ heater is used with a 50 HP, 230 V, 3φ motor.

_____ 8. A number ___ heater is used with a 15 HP, 230 V, 3φ motor with a size 2 starter.

_____ 9. A number ___ heater is used with a 20 HP, 460 V, 3φ motor with a size 2 starter.

_____ 10. A number ___ heater is used with a ¾ HP, 230 V, 3φ motor with a size 0 starter.

Activity 8-4: Contactor and Motor Starter Ratings

Determine the correct contactor or motor starter for each application using Control Ratings in Appendix.

_____ 1. A size ___ motor starter is used to switch a 1.5 HP, 230 V, 1φ (normal duty) motor.

_____ 2. A size ___ motor starter is used to switch a 115 V, 1φ, 41 A (continuous duty) motor.

_____ 3. A size ___ motor starter is used to switch a 27 HP, 460 V, 3φ (normal duty) motor.

_____ 4. A size ___ motor starter is used to switch a 27 HP, 460 V, 3φ (plugging duty) motor.

_____ 5. A size ___ motor starter is used to switch a 460 V, 3φ, 125 A (continuous duty) motor.

_____ 6. A size ___ contactor is used to switch a 230 V, 50 A tungsten lamp.

_____ 7. A size ___ contactor is used to switch a 115 V, 1φ, 50 A heating element.

_____ 8. A size ___ contactor is used to switch a 230 V, 1φ, 50 A heating element.

_____ 9. A size ___ contactor is used to switch (less than 20 times per hour) a 460 V, 3φ, 5 kVA transformer.

_____ 10. A size ___ contactor is used to switch a 230 V, 10 kVA capacitor bank.

Activity 8-5: Sizing Motor Protection, Motor Starters, and Wire

List the devices based on the motor installation data. Use 1φ Motors and Circuits in Appendix.

A 1 HP, 230 V motor with a 40°C rating and a 1.15 SF is installed in a 60°C (or less) location.

_____ 1. The motor overload protection fuse is ___ A.

_____ 2. The motor starter size is ___.

_____ 3. The wire size is No. ___.

_____ 4. The conduit size is ___″.

A ½ HP, 230 V motor with a 40°C rating and a 1.15 SF is installed in a 60°C (or less) location.

_____ 5. The motor overload protection fuse is ___ A.

_____ 6. The motor starter size is ___.

_____ 7. The wire size is No. ___.

_____ 8. The conduit size is ___″.

A ¾ HP, 115 V motor with a 40°C rating and a 1.15 SF is installed in a 60°C (or less) location.

_____ 9. The motor overload protection fuse is ___ A.

_____ 10. The motor starter size is ___.

_____ 11. The wire size is No. ___.

_____ 12. The conduit size is ___".

A 2 HP, 115 V motor with a 50°C rating and a 1.15 SF is installed in a 60°C (or less) location.

_____ 13. The motor overload protection fuse is ___ A.

_____ 14. The motor starter size is ___.

_____ 15. The wire size is No. ___.

_____ 16. The conduit size is ___".

List the devices based on the motor installation data. Use 3φ Motors and Circuits in Appendix.
A 3 HP, 230 V motor with a 40°C rating and a 1.15 SF is installed in a 60°C (or less) location.

_____ 17. The motor overload protection fuse is ___ A.

_____ 18. The motor starter size is ___.

_____ 19. The wire size is No. ___.

_____ 20. The conduit size is ___".

A 75 HP, 230 V motor with a 40°C rating and a 1.15 SF is installed in a 60°C (or less) location.

_____ 21. The motor overload protection fuse is ___ A.

_____ 22. The motor starter size is ___.

_____ 23. The wire size is No. ___.

_____ 24. The conduit size is ___".

A 40 HP, 460 V motor with a 40°C rating and a 1.15 SF is installed in a 60°C (or less) location.

_____ 25. The motor overload protection fuse is ___ A.

_____ 26. The motor starter size is ___.

_____ 27. The wire size is No. ___.

_____ 28. The conduit size is ___".

List the devices based on the motor installation data. Use DC Motors and Circuits in Appendix.

A 1 HP, 120 V motor with a 40°C rating and a 1.15 SF is installed in a 60°C (or less) location.

_____ 29. The motor overload protection fuse is ___ A.

_____ 30. The motor starter size is ___.

_____ 31. The wire size is No. ___.

_____ 32. The conduit size is ___″.

A 10 HP, 120 V motor with a 40°C rating and a 1.15 SF is installed in a 60°C (or less) location.

_____ 33. The motor overload protection fuse is ___ A.

_____ 34. The motor starter size is ___.

_____ 35. The wire size is No. ___.

_____ 36. The conduit size is ___″.

A 3 HP, 180 V motor with a 40°C rating and a 1.15 SF is installed in a 60°C (or less) location.

_____ 37. The motor overload protection fuse is ___ A.

_____ 38. The motor starter size is ___.

_____ 39. The wire size is No. ___.

_____ 40. The conduit size is ___″.

Activity 8-6: Motor Starter Replacement Parts

Answer the questions using the Service Bulletin on page 95.

_____ 1. The part number used to replace a normally closed internal interlock is ___.

_____ 2. The part number of the movable contact is ___.

_____ 3. The part number of the clamp that holds the incoming power lines and motor terminal wires is ___.

_____ 4. How many clamps are required for the starter when controlling a 3ϕ motor?

_____ 5. What is the difference between items B and C?

Activity 8-7: Motor Drives

1. Connect the motor to the drive power terminals for low voltage. Connect the drive power terminals to the main disconnect. Connect the drive for operation using an external forward and reverse pushbutton, and using an external potentiometer for speed control.

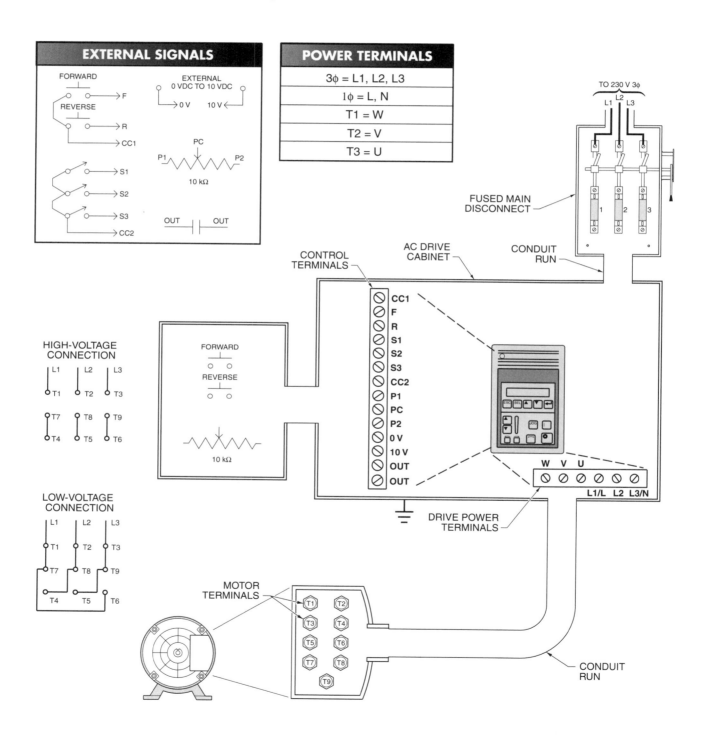

2. Connect the motor to the drive power terminals for high voltage. Connect the drive power terminals to the main disconnect. Connect the drive for operation using an external selector switch so the low position sends a signal to S1 and the high position sends a signal to S2. Connect the drive using an external 0 VDC to 10 VDC power source for speed control.

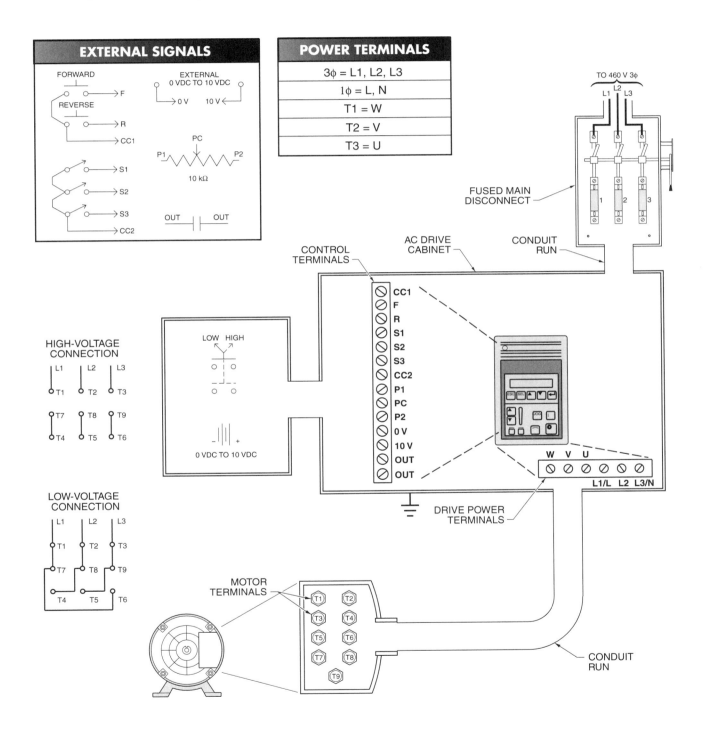

Activity 8-8: Troubleshooting Contactors

1. Connect DMM 1 so the meter reads the primary voltage at the control transformer. Connect DMM 2 so the meter reads the voltage at the heating contactor. Connect DMM 3 so the meter checks the voltage delivered from the temperature switch when the switch is closed. Connect DMM 4 so the meter reads the voltage applied to the input of the temperature switch.

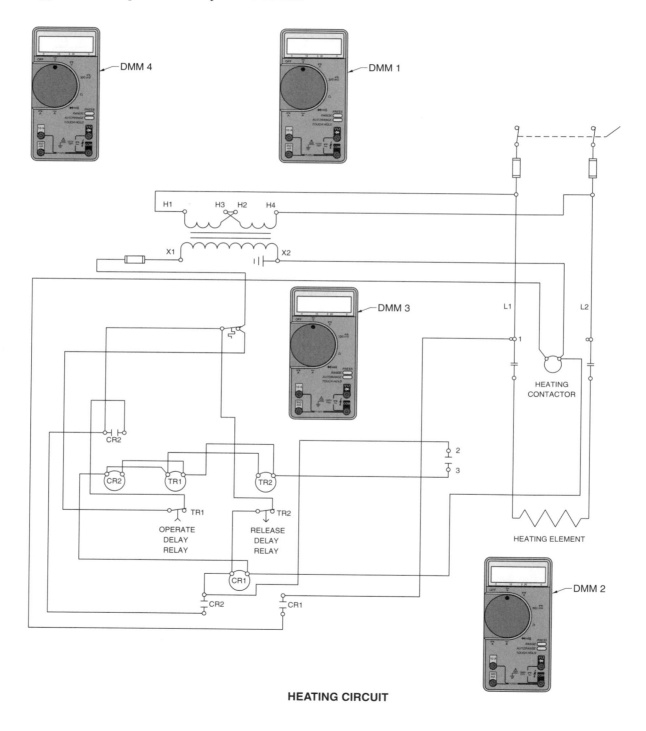

HEATING CIRCUIT

When troubleshooting heating circuits, approximate meter readings should be anticipated if the meter readings are going to be used to help determine circuit problems. Determine the expected DMM readings if the circuit is operating properly.

_____ 2. The expected reading of DMM 1 with the heating element OFF is ___ VAC.

_____ 3. The expected reading of DMM 1 with the heating element ON is ___ VAC.

_____ 4. The expected reading of DMM 2 with the heating element OFF is ___ VAC.

_____ 5. The expected reading of DMM 2 with the heating element ON is ___ VAC.

_____ 6. The expected reading of the clamp-on ammeter with the heating element ON is ___ A.

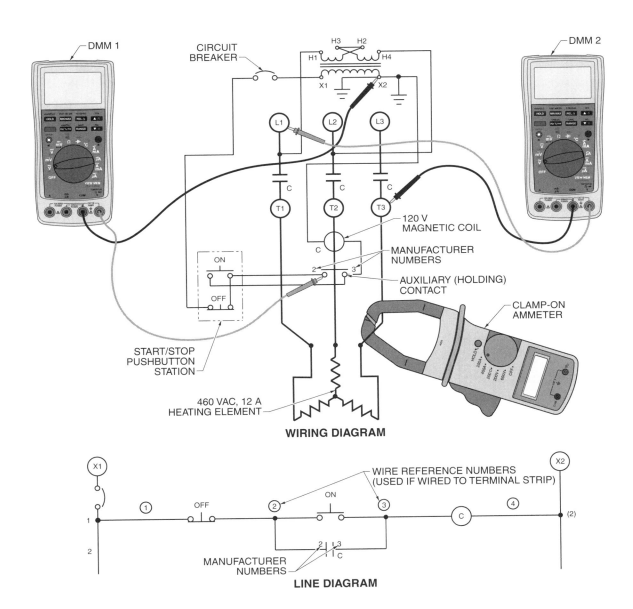

When troubleshooting heating power and control circuits, approximate meter readings should be anticipated if the meter readings are going to be used to help determine circuit problems. Determine the DMM expected readings if the circuit is operating properly.

_____ 7. The expected reading of DMM 1 with the heater OFF is ___ VAC.

_____ 8. The expected reading of DMM 1 with the heater ON is ___ VAC.

_____ 9. The expected reading of DMM 2 with the selector switch ON and the heater OFF is ___ VAC.

_____ 10. The expected reading of DMM 2 with the selector switch ON and the heater ON is ___ VAC.

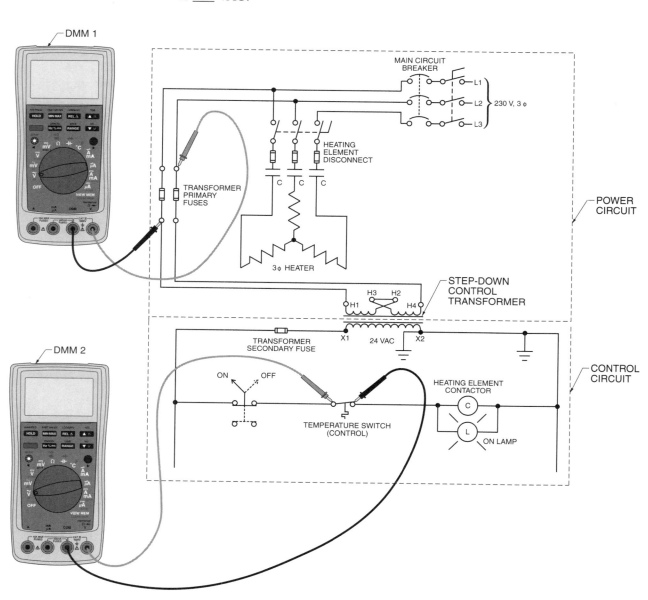

Control Devices Applications

Application 9-1: Enclosure Selection

Enclosures

An *enclosure* is a device that protects a motor starter and control devices. Enclosures are categorized by the protection they provide. An enclosure is selected based on the location of the equipment and the National Electrical Code® (NEC®) requirements. **See NEMA Enclosure Classification.** For example, a Type 3 enclosure is used outdoors for applications subjected to windblown dust, rain, sleet, and ice. Type 3 enclosures must pass rain, external icing, windblown dust, and rust resistance tests. Type 3 enclosures, however, do not protect against internal condensation or internal icing.

Enclosures are typically constructed of cold rolled steel for indoor use and galvanized steel for indoor and outdoor use. Knockouts are provided to allow installation of conduit and cable fittings. Knockouts can be removed prior to mounting or after the enclosure is mounted. Enclosures may contain single-ring and/or multiple-ring knockouts.

To remove a single-ring knockout, the knockout is struck at the point furthest from the tie (place where the knockout is connected to the enclosure) and bent back and forth to break the tie. To remove a multiple-ring knockout, the small center knockout is removed the same way as a single-ring knockout. If a larger opening is required, each additional ring is removed one at a time by prying it with a screwdriver and bending it back and forth with pliers. **See Single-Ring Knockout and Multiple-Ring Knockout.**

NEMA ENCLOSURE CLASSIFICATION

Type	Use	Service Conditions	Tests	Comments
1	Indoor	No unusual	Rod entry, rust resistance	
3	Outdoor	Windblown dust, rain, sleet, and ice on enclosure	Rain, external icing, dust, and rust resistance	Does not provide protection against internal condensation or internal icing
3R	Outdoor	Falling rain and ice on enclosure	Rod entry, rain, external icing, and rust resistance	Does not provide protection against dust, internal condensation, or internal icing
4	Indoor/outdoor	Windblown dust and rain, splashing water, hose-directed water, and ice on enclosure	Hosedown, external icing, and rust resistance	Does not provide protection against internal condensation or internal icing
4X	Indoor/outdoor	Corrosion, windblown dust and rain, splashing water, hose-directed water, and ice on enclosure	Hosedown, external icing, and corrosion resistance	Does not provide protection against internal condensation or internal icing
6	Indoor/outdoor	Occasional temporary submersion at a limited depth		
6P	Indoor/outdoor	Prolonged submersion at a limited depth		
7	Indoor locations classified as Class I, Groups A, B, C, or D, as defined in the NEC®	Withstand and contain an internal explosion of specified gases, contain an explosion of specified gases, contain an explosion sufficiently so an explosive gas-air mixture in the atmosphere is not ignited	Explosion, hydrostatic, and temperature	Enclosed heat-generating devices shall not cause external surfaces to reach temperatures capable of igniting explosive gas-air mixtures in the atmosphere
9	Indoor locations classified as Class II, Groups E or G, as defined in the NEC®	Dust	Dust penetration, temperature, and gasket aging	Enclosed heat-generating devices shall not cause external surfaces to reach temperatures capable of igniting explosive gas-air mixtures in the atmosphere
12	Indoor	Dust, falling dirt, and dripping noncorrosive liquids	Drip, dust, and rust resistance	Does not provide protection against internal condensation
13	Indoor	Dust, spraying water, oil, and noncorrosive coolant	Oil explosion and rust resistance	Does not provide protection against internal condensation

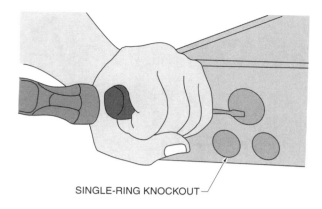

SINGLE-RING KNOCKOUT

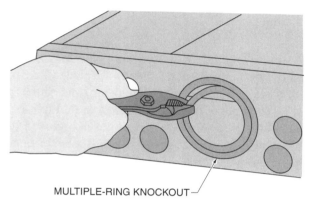

MULTIPLE-RING KNOCKOUT

SINGLE-RING KNOCKOUT AND MULTIPLE-RING KNOCKOUT

Hazardous Locations

The National Electrical Code® (NEC®) classifies hazardous locations according to the properties and quantities of the hazardous material that may be present. Hazardous locations are divided into three classes, two divisions, and seven groups. *Class* refers to the generic hazardous material present. Class I applies to locations where flammable gases or vapors may be present in the air in quantities sufficient to produce an explosive or ignitable mixture. Class II applies to locations where combustible dusts may be present in sufficient quantity to cause an explosion. Class III applies to locations where the hazardous material consists of easily ignitable fibers or flyings that are not normally in suspension in the air in large enough quantities to produce an ignitable mixture.

Division applies to the probability that a hazardous material is present. Division 1 applies to locations where ignitable mixtures exist under normal operating conditions found in the process, in the operation, or during periodic maintenance. Division 2 applies to locations where ignitable mixtures exist only in abnormal situations. Abnormal situations occur as a result of accidents or when equipment fails.

A *group* is an atmosphere containing flammable gases or vapors or combustible dust. Air mixtures of gases, vapors, and dusts are grouped according to their similar characteristics. The NEC® classifies gases and vapors in Groups A, B, C, and D for Class I locations and combustible dusts in Groups E, F, and G for Class II locations. **See Hazardous Locations.** For example, a Type 7 enclosure is required for an indoor application where gasoline is stored (Class I, Group D).

Class	Group	Material
		HAZARDOUS LOCATIONS
I	A	Acetylene
	B	Hydrogen, butadiene, ethylene oxide, propylene oxide
	C	Carbon monoxide, ether, ethylene, hydrogen sulfide, morpholine, cyclopropane
	D	Gasoline, benzene, butane, propane, alcohol, acetone, ammonia, vinyl chloride
II	E	Metal dusts
	F	Carbon black, coke dust, coal
	G	Grain dust, flour, starch, sugar, plastics
III	No Groups	Wood chips, cotton, flax, nylon

Application 9-2: Alternating Motor Control

Dual-Motor Circuits

Using dual motors in an application prevents downtime and reduced production. The two main advantages of using dual motors in an application are that one motor can do work if the other motor is down, and both motors can be energized if extra work is required. Another advantage of dual-motor circuits is that maintenance schedules can be alternated for each motor, minimizing standard preventive maintenance downtime.

The three control options possible with dual-motor circuits are as follows:

- One motor (the main motor) is used each time work is required. The other motor (the backup motor) is used when a problem occurs with the main motor. The disadvantage of this control option is that one motor is always working and the other motor is always idle, which is not good for either motor.
- One motor is used for a time period (a day or a week) and the other motor remains idle. After the time period, the motor functions are switched. The disadvantage of this control option is that a rotation schedule is required for each motor.
- The motors are alternated from rest to work each time work is required. The motor alternation is accomplished through an alternating (flip-flop) relay. The advantage of this control option is that both motors are ON approximately the same amount of time over the long run. This is the best control option for most dual-motor applications.

Understanding the logic of a control circuit requires understanding how each component operates, and how numbering systems are used on manufacturer specifications. Certain components, such as pushbuttons and level switches, are less difficult to understand because they are either normally open or normally closed, depending on circuit conditions. Understanding the operation of other components, such as a flip-flop relay, requires understanding the manufacturer specifications for the device. Manufacturer specifications include the relay operational diagram that shows how the component operates under different circuit conditions, the actual numbers of the pins (terminals) that must be wired into the circuit from the component, and written guidelines. Understanding all information is required when servicing or troubleshooting a circuit. **See Flip-Flop Relay.**

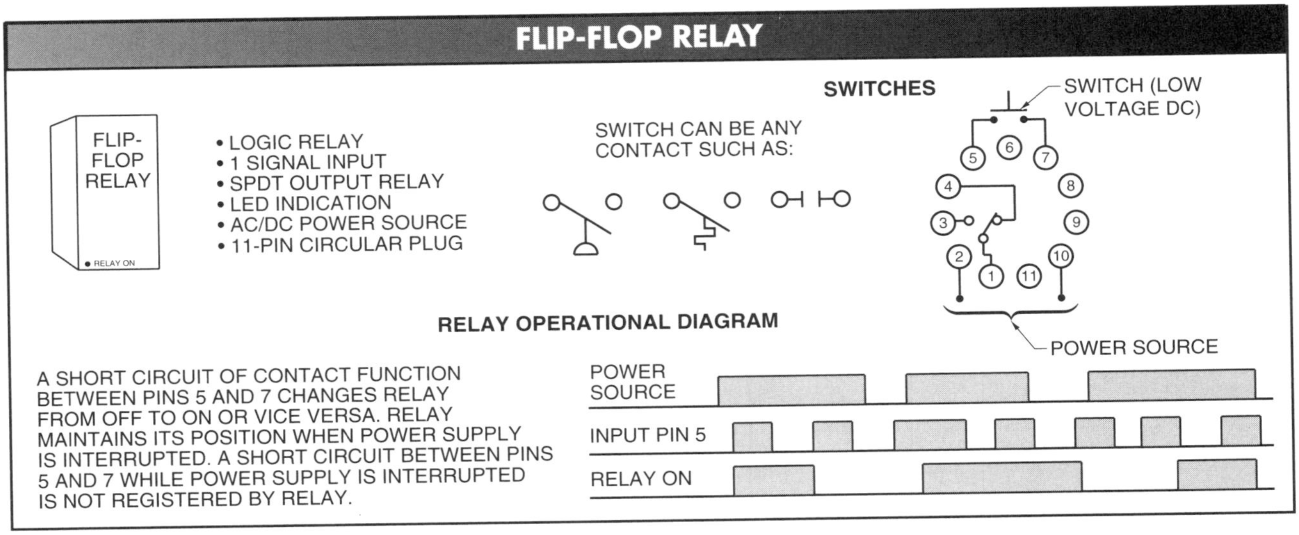

FLIP-FLOP RELAY
FLIP-FLOP RELAY
RELAY ON
• LOGIC RELAY
• 1 SIGNAL INPUT
• SPDT OUTPUT RELAY
• LED INDICATION
• AC/DC POWER SOURCE
• 11-PIN CIRCULAR PLUG
SWITCHES
SWITCH CAN BE ANY CONTACT SUCH AS:
SWITCH (LOW VOLTAGE DC)
POWER SOURCE
RELAY OPERATIONAL DIAGRAM
A SHORT CIRCUIT OF CONTACT FUNCTION BETWEEN PINS 5 AND 7 CHANGES RELAY FROM OFF TO ON OR VICE VERSA. RELAY MAINTAINS ITS POSITION WHEN POWER SUPPLY IS INTERRUPTED. A SHORT CIRCUIT BETWEEN PINS 5 AND 7 WHILE POWER SUPPLY IS INTERRUPTED IS NOT REGISTERED BY RELAY.
POWER SOURCE
INPUT PIN 5
RELAY ON

Application 9-3: Level Control

Charging and Discharging

Level controls are relays that maintain a set level in a tank. A level control keeps a tank full or empty. *Charging* is the process of filling a tank or keeping a tank full. Charging a tank is accomplished by a solenoid valve or pump motor. A solenoid valve is used in applications where the tank to be charged is located at a lower elevation than the supply tank, or in an application where the product to be placed in the tank to be charged is under pressure. A pump motor is used in applications where the tank to be charged is level with or above the supply tank.

Discharging is the process of emptying a tank. Like charging, discharging a tank is performed by a solenoid valve or pump motor. Whether a solenoid valve or pump motor is used for discharging depends on the viscosity of the product, the location of the tank to be charged, and the distance the product must be moved.

A level-control relay is used to control the charging or discharging of a tank. A level-control relay controls a pump motor or valve. Level probes that detect the product inside the tank are connected to the relay. Level-control relays are typically used in coating, chemical, food, and wastewater treatment industries. **See Level-Control Relay.**

A level-control relay is used to control the level of product, simplify installation, and add safety features to an application. For example, level-control relay includes a step-down transformer that changes the high control circuit voltage (L1/X1 and L2/X2) to a lower voltage and a rectifier circuit that changes the low AC voltage to DC voltage at the pins of the relay that are connected to detect the liquid level in the tank (manufacturer pins 5, 6, and 7). Understanding that a control circuit can include different voltage types (AC and DC) and different voltage levels (12 V, 24 V, 120 V, etc.) is required when troubleshooting a circuit.

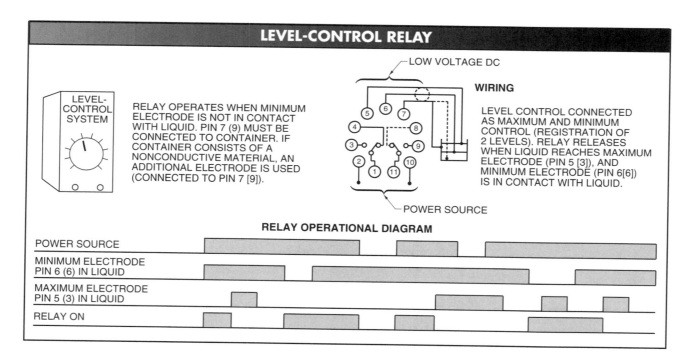

Application 9-4: Temperature Control

Temperature-Control Relays

Temperature controls are temperature-sensitive relays that maintain the proper temperature of an application. A temperature switch may be used for heating or cooling applications, depending on whether the contacts are normally open or normally closed. With some temperature controls, cooling or heating control is determined by a switch or interconnecting pins. **See Temperature-Control Relay.**

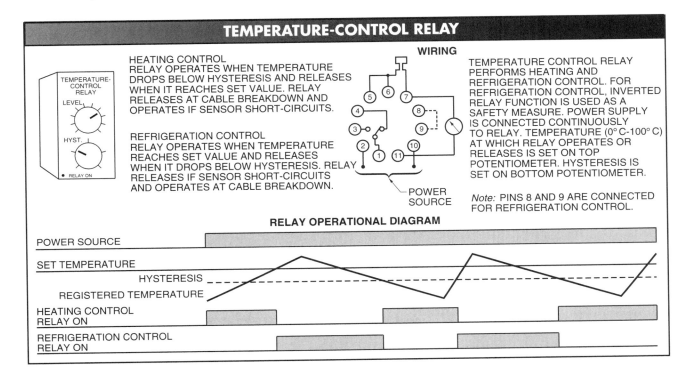

Application 9-5: Selecting Blowers and Exhaust Fans

Airflow Rate

Airflow rate is the volumetric flow rate of air. Proper airflow rate is required for many commercial and industrial applications. Blowers and fans are used to circulate, move, and exhaust air through HVAC systems. Blowers and fans are rated according to their blade diameter for fans (18″, 24″, etc.), wheel diameter for blowers (9″, 12″ etc.), motor horsepower (¼ HP, 1 HP, etc.), speed (690 rpm, 1140 rpm, etc.), voltage requirements (115 VAC, 230 VAC, 460 VAC, etc.), wattage, and volume of output airflow in cubic feet per minute (cfm). The greater the volume of output airflow, the higher the blower/fan cost and operating cost.

The amount of airflow depends on the application and required amount of ventilation air. **See Required Air Change/Exhaust Ventilation Guide.** Required air change/exhaust ventilation guides list the range of number of air changes required for a given area (application). The guides also list a typical selection factor which is a common factor used for the area. The amount of airflow required for an application is found by applying the following formula:

$$Q = \frac{V \times R_{AC}}{60}$$

where

Q = required airflow rate (in cfm)
V = room volume (in cu ft)
R_{AC} = required number of air changes per hour
60 = constant

> **Example: Determining Airflow Rate**
> What is the airflow rate for a cafeteria that is $100' \times 80' \times 10'$ using a typical selection factor?
>
> $$Q = \frac{V \times R_{AC}}{60}$$
> $$Q = \frac{(100 \times 80 \times 10) \times 4}{60}$$
> $$Q = \frac{80,000 \times 4}{60}$$
> $$Q = \frac{320,000}{60}$$
> $$Q = \mathbf{5333 \text{ cfm}}$$
>
> The airflow rate for a $100' \times 80' \times 10'$ cafeteria is 5333 cfm.

REQUIRED AIR CHANGE/EXHAUST VENTILATION GUIDE

Area	Required Air Changes per Hour	Typical Selection Factor
Auditoriums/Theaters (small capacity)	1-5	4
Auditoriums/Theaters (normal capacity)	2-10	6
Auditoriums/Theaters (large capacity)	4-20	8
Bakeries	1-3	2
Banks	2-8	5
Bars	2-6	4
Cafeterias	3-5	4
Classrooms	3-8	5
Dormitories/Hotel Rooms	3-8	5
Factories (clean assembly areas)	3-8	5
Factories (dirty production areas)	4-10	7
Garages	4-10	6
Gymnasiums	2-10	5
Hallways (lightly used)	2-6	3
Hallways (heavily used)	4-12	6
Kitchens (commercial)	1-5	3
Libraries	2-5	3
Locker Rooms (lightly used)	2-6	4
Locker Rooms (heavily used)	4-10	6
Markets (retail stores)	2-8	5
Restaurants	2-8	5
Warehouses (storage)	2-6	3
Washrooms (lightly used)	3-8	5
Washrooms (heavily used)	4-12	7

Selecting Blowers and Exhaust Fans

The proper blower or fan must be selected for an application. Blowers and exhaust fans may be selected for an application based on the blade diameter (for fans), wheel diameter (for blowers), motor horsepower, current draw, voltage requirements, and/or required airflow rate. **See Blower/Fan Specifications.**

BLOWER/FAN SPECIFICATIONS

Model	Blade Diameter*	Power†	Current Draw‡			Air Flow§
			115 VAC	230 VAC	460 VAC	
12-1	12	¼	4	1.2	NA	1600
12-2	12	¼	NA	1.2	0.6	1600
24-1	24	⅓	5	1.9	NA	3270
24-2	24	⅓	NA	1.9	0.9	3270
30-1	30	⅓	6	2.1	NA	7730
30-2	30	⅓	NA	2.1	1	7730
36-1	36	½	8	2.3	NA	9870
36-2	36	½	NA	2.3	1.1	9870
42-1	42	¾	11	3.3	NA	14,800
42-2	42	¾	NA	3.3	1.5	14,800
48-1	48	1	15	3.8	NA	21,500
48-2	48	1	NA	3.8	1.9	21,500

* in in.
† in HP
‡ in A
§ in cfm

NA = Not Applicable

Example: Selecting Blowers and Exhaust Fans

What is the minimum size blower required for a cafeteria application that requires 5333 cfm using two fans of the same size with a 115 VAC power source?

Two identical fan models must have a total airflow rate of at least 5333 cfm. For this reason, each fan must produce at least 2666.5 cfm. Using the Blower/Fan Specifications chart, the minimum model that fits this requirement and also operates on 115 VAC is Model 30-1. Two Model 30-1 units are the minimum required for the cafeteria application.

Application 9-6: Troubleshooting Control Device Circuits

Testing Switches

When troubleshooting control device circuits, the function of the circuit and expected voltage from a properly operating circuit must be known. A deviation from the expected voltage indicates a fault in the tested part of the circuit. A suspected fault with an electromechanical switch is tested using a DMM set to measure voltage. The voltage setting on the DMM is used to test the voltage flowing into and out of the switch.

If more than one control switch is used to control a load (motor starter, solenoid, etc.), then preliminary troubleshooting should be performed prior to connecting meters into the circuit. For example, when a selector switch is used to place a circuit in the HAND or AUTO operating mode, the circuit is tested to see if either or both circuit conditions are operating. If the motor turns ON in either the HAND or AUTO operating mode, there is no deficiency with the control transformer, fuses/circuit breakers/overloads, motor, or motor starter. If the motor does not turn ON in either the HAND or AUTO operating mode, then the deficiency is located in the power circuit or control circuit. A voltage measurement taken at the primary and secondary sides of the control circuit transformer determines whether the problem is in the power circuit (no primary voltage is present) or the control circuit (voltage is present at the secondary side, but the control circuit fails to operate).

Measuring the voltage into and out of a control switch (pushbutton, selector switch, daylight switch, etc.) can determine if a switch is operating properly but can be difficult if the switch's measurement parameters are not met. For example, a pressure switch must be at the set amount of force to operate its contacts, a temperature switch must be at the set heat intensity to operate its contacts, and a daylight switch will not close its contacts until there is a change in light intensity. When checking a daylight switch, the switch's light sensor can be covered to simulate darkness. When checking a pressure switch or temperature switch, a fused jumper is used to bypass the switch so that the entire circuit can be checked.

When checking the voltage into and out of a switch, it is important to monitor the two voltage measurements. In theory, the voltage into a switch should be the same as the voltage out of the switch. However, there is almost always a slight voltage drop between the voltage into a switch and the voltage out of the switch. The higher the voltage drop, the higher the resistance of the switch contacts (meaning that the contacts are failing). The voltage drop across mechanical switches should be less than 0.25% of the total applied voltage. For example, since power equals voltage times current ($P = E \times I$), a voltage drop of 2 V with 5 A flowing through a switch produces 10 W of power loss (heat) at the switch. This heat shortens the life of the switch.

Control Devices

Activities 9

Name _____ Date _____

Activity 9-1: Enclosure Selection

List the enclosure that best fits the application using the NEMA Enclosure Classification chart on page 107 and the Hazardous Locations chart on page 108.

_____ 1. A Type ___ enclosure is used for a commercial basement that occasionally floods.

_____ 2. A Type ___ enclosure is used for a warehouse used to store paper.

_____ 3. A Type ___ enclosure is used in an industrial Class II, Group E location.

_____ 4. A Type ___ enclosure is used in an industrial Class I, Group B location.

_____ 5. A Type ___ enclosure is used in an industrial building in which the enclosure is occasionally hosed down with water to remove a noncorrosive coolant.

_____ 6. A Class ___ rating applies to locations where wood chips are present.

_____ 7. A Class ___ rating applies to locations where flammable gases or vapors are present in quantities sufficient to produce an explosive or ignitable mixture.

_____ 8. A Class ___ rating applies to locations where combustible dusts may be present in sufficient quantity to produce an explosion.

_____ 9. Division ___ applies to areas in which a hazardous material is probably present.

_____ 10. Division ___ applies to areas in which a hazardous material may be present in abnormal situations.

Activity 9-2: Alternating Motor Control

Answer the questions using the Flip-Flop Relay on page 109 and Alternating Motor Control Circuit.

_____ 1. Which electrical input device turns a pump motor ON or OFF?

_____ 2. Can motor 1 and motor 2 ever be ON at the same time?

_____ 3. If motor 1 is running and its overload trips (opens), does motor 2 (M2) automatically turn on?

4. If there is the proper voltage measured between wire reference numbers 3 and 2, but no measurable voltage between wire reference numbers 4 and 2, the problem is most likely caused by ___ .

5. If lamp 1 (L1) is connected into the circuit so that it is ON every time pump 1 is ON, the lamp is connected to which two wire reference numbers?

6. If lamp 2 (L2) is connected into the circuit so that it is ON every time pump 2 is ON, the lamp is connected to which two wire reference numbers?

7. If lamp 1 (L1) is connected into the circuit so that it is ON every time pump 2 is ON, the lamp is connected to which two wire reference numbers?

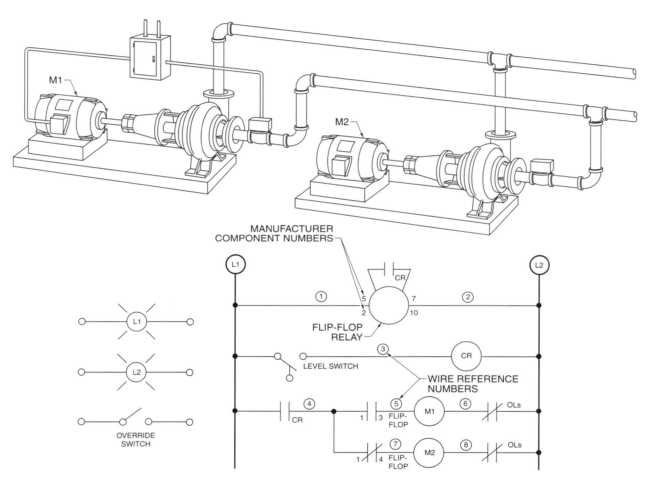

ALTERNATING MOTOR CONTROL CIRCUIT

Activity 9-3: Level Control

1. Draw the correct position of the selector switch on DMM 1 to measure voltage at pin numbers 2 and 10. Draw the correct position of the selector switch on DMM 2 to measure voltage at pin numbers 5, 6, and 7. Connect DMM 1 to the power supply input of the level-control relay. Connect DMM 2 to the high-level output pin and pin number 7 of the level-control relay.

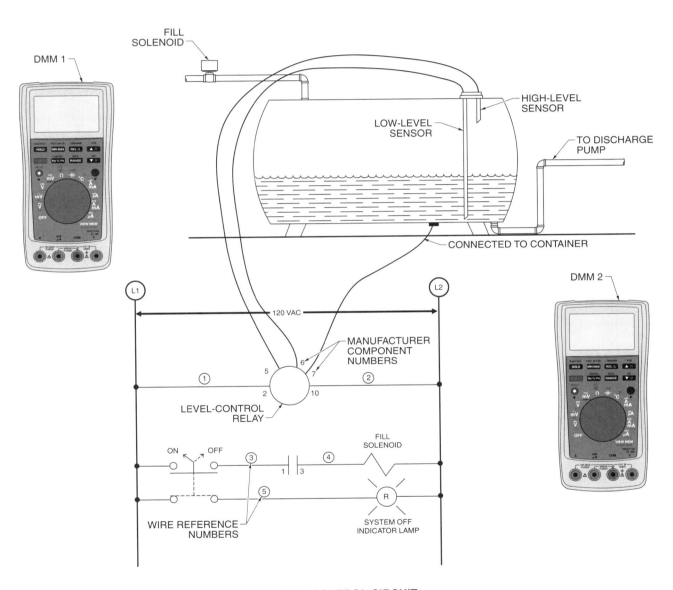

LEVEL-CONTROL CIRCUIT

Activity 9-4: Temperature Control

Logic relays can be used together to develop complex control functions. In the Temperature Control Circuit, a level-control relay and temperature-control relay are used together. Answer the questions using the Temperature-Control Circuit, the Level-Control Relay on page 110, and Temperature-Control Relay on page 111.

_____ 1. Is it possible for the heating element to be ON if the level in the tank is below the high-level sensor probe but higher than the low-level sensor probe?

_____ 2. Is it possible for the heating element to be ON if the level in the tank is below the low-level sensor probe?

_____ 3. Is it possible for the heating element to be ON in any circuit condition if the temperature-control relay has detected that the temperature is above the setting of the temperature-control relay?

4. How does connecting manufacturer component pins 8 and 9 together on the temperature-control relay change the relay operation?

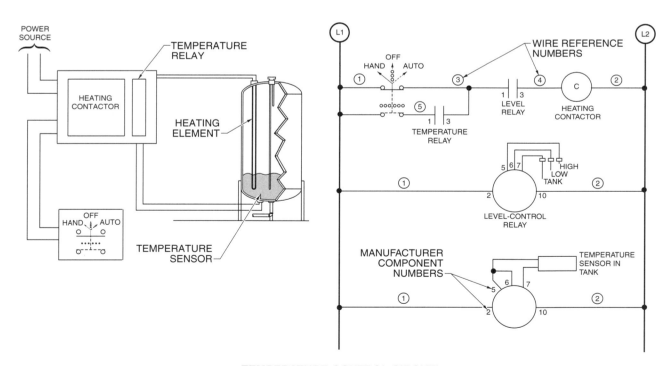

TEMPERATURE-CONTROL CIRCUIT

Activity 9-5: Selecting Blowers and Exhaust Fans

Using the Required Air Change/Exhaust Ventilation Guide on page 112 and the Blower/Fan Specifications on page 113, determine the required blower/fan airflow rating and model number for each application.

_____ 1. The required blower/fan airflow rating for a heavily used washroom that is 60′ × 20′ × 10′, using a typical selection factor, is ___ cfm.

_____ 2. The minimum circulating/exhaust fan required for the heavily used washroom application using one fan with a 115 VAC power source is Model ___.

_____ 3. The minimum circulating/exhaust fan required for the heavily used washroom application using one fan with a 460 VAC power source is Model ___.

_____ 4. The required blower/fan airflow rating for a normal theater that is 250′ × 90′ × 24′, using a typical selection factor, is ___ cfm.

_____ 5. The minimum circulating/exhaust fan required for the normal theater application using six fans of the same size with a 115 VAC power source is Model ___.

_____ 6. The minimum circulating/exhaust fan required for the normal theater application using six fans of the same size with a 460 VAC power source is Model ___.

_____ 7. A Model 30-1 blower/fan motor can be connected to what two different voltage levels?

_____ 8. What is the advantage of connecting the Model 30-1 blower/fan motor for the higher voltage rating?

_____ 9. If an application requires an airflow of 3270 cfm, what motor voltage ratings are available?

_____ 10. If 115 VAC is the only easily available voltage, what is the highest airflow (in cfm) blower/fan model available?

Activity 9-6: Troubleshooting Control Device Circuits

Answer the questions using the Selector Switch Circuit. Note: *The motor does not operate in the AUTO or OFF position. The motor does operate in the HAND position.*

_____ 1. The ___ is the most likely cause of the malfunction.

_____ 2. Is it likely that the problem is in the overloads?

_____ 3. Is it likely that the problem is in the motor starter coil?

_____ 4. Is it likely that the problem is in the selector switch?

_____ 5. Is it likely that the problem is in the pressure switch?

_____ 6. Is it likely that the problem is in the motor?

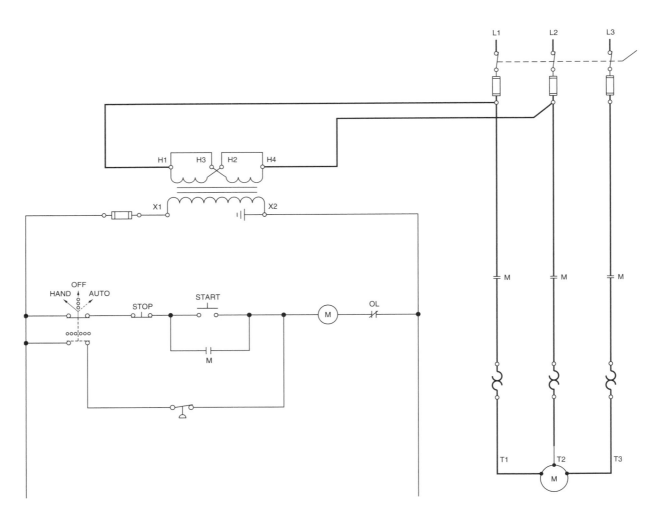

SELECTOR SWITCH CIRCUIT

When troubleshooting daylight switch circuits, approximate meter readings should be anticipated if the meter readings are going to be used to help determine circuit problems. Determine the expected DMM readings if the circuit is operating properly.

_____ 7. The expected daytime reading on DMM 1 is ___ VAC.

_____ 8. The expected nighttime reading on DMM 1 is ___ VAC.

_____ 9. The expected daytime reading on DMM 2 is ___ VAC.

_____ 10. The expected nighttime reading on DMM 2 is ___ VAC.

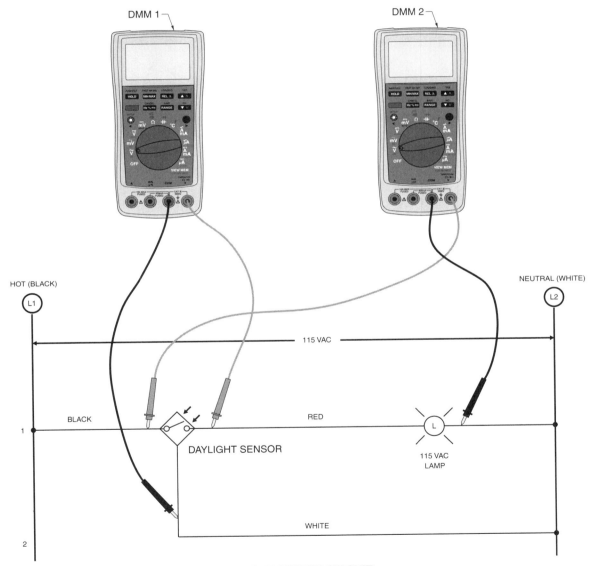

DAYLIGHT SWITCH CIRCUIT

When troubleshooting HVAC circuits, approximate meter readings should be anticipated if the meter readings are going to be used to help determine circuit problems. Determine the expected DMM readings if the circuit is operating properly.

_____ 11. The expected reading of DMM 1 with the air conditioning unit ON is ___ VAC.

_____ 12. The expected reading of DMM 1 with the air conditioning unit OFF is ___ VAC.

_____ 13. The expected reading of DMM 2 with the air conditioning unit ON is ___ VAC.

_____ 14. The expected reading of DMM 2 with the air conditioning unit OFF is ___ VAC.

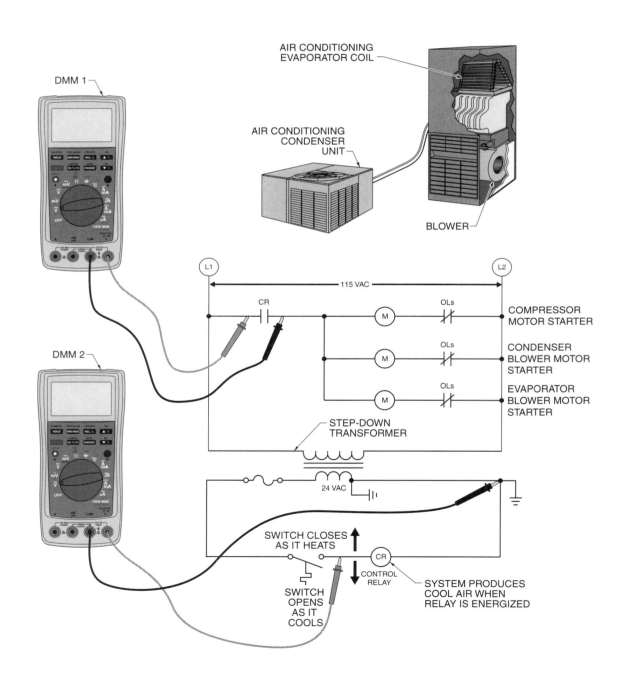

Reversing Motor Circuits

Applications 10

Application 10-1: Reversing Motor Circuits

Reversing Motors

Motors are designed to rotate in the clockwise or counterclockwise direction. Clockwise (FWD) or counterclockwise (REV) rotation is determined by viewing the front of a motor. Normally, the front of a motor is the end opposite the shaft.

Any motor that is designed to rotate in both directions can be reversed. Reversing a motor is accomplished by interchanging motor leads. The motor leads are connected to a reversing drum switch, manual reversing starter, or magnetic reversing starter. **See Reversing Methods.**

Reversing Drum Switches

A *reversing drum switch* is a three-position manual switch that has six terminal connections. The motor leads are connected to the terminal connections. A drum switch interchanges the motor leads and does not provide overload protection. Drum switches are normally used with fractional horsepower motors that control machine tools, cranes, and forklifts.

Manual Reversing Starters

A *manual reversing starter* is a switch that includes overload protection. Overload protection protects the motor when running and automatically disconnects an overloaded motor. To prevent a short circuit and damage to a motor, the starter provides mechanical interlocking. Mechanical interlocking prevents the forward and reverse contacts from being energized at the same time. Manual reversing starters are normally used with motors less than 2 HP.

Magnetic Reversing Starters

A *magnetic reversing starter* is a switch that uses starting coils to control the position of the contacts. By using starting coils, the starter can be remotely controlled. The forward coil energizes the forward set

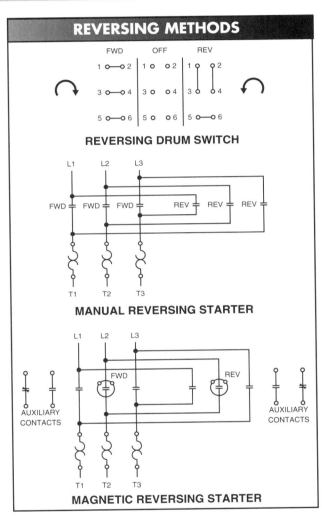

of contacts, and the reverse coil energizes the reverse set of contacts. A magnetic reversing starter provides mechanical interlocking. Magnetic reversing starters are used with all sizes of motors and are the most common starter used in industrial applications. Magnetic reversing starters include overload protection.

Note: Wire numbering systems vary by motor manufacturer. Wire numbers may be given on the nameplate, on the wires, or near terminal connections.

Application 10-2: Reversing Three-Phase Motors

Three-Phase Motors

Three-phase (3ϕ) motors are reversed by interchanging two of the three main power lines of the motor. The industry standard is to interchange line one (L1) and line three (L3). The motor is connected L1 to T1, L2 to T2, and L3 to T3 for one direction of rotation, and L3 to T1, L2 to T2, and L1 to T3 for the opposite direction of rotation. **See Three-Phase Motor Wiring Diagrams.**

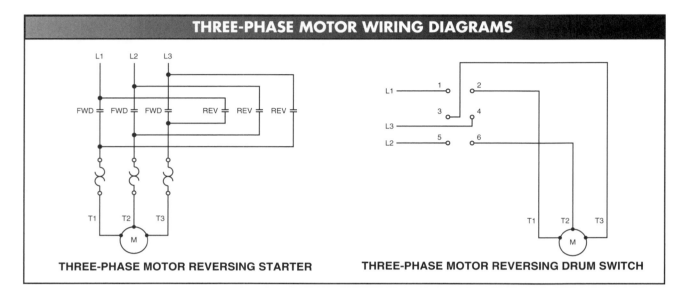

Application 10-3: Reversing Single-Phase Motors

Single-Phase Motors

A single-phase (1ϕ) motor requires a starting winding for starting. The starting winding gives the motor starting torque and determines the direction of motor rotation. A 1ϕ motor is reversed by interchanging the leads of the starting or running windings. A motor starter is used to reverse the starting windings. The motor starter is used to interchange the starting winding leads and to disconnect the running windings from power. **See Single-Phase Motor Wiring Diagrams.**

Capacitor-Start Motors

Capacitor-start motors have a capacitor added in series with the starting winding of a 1ϕ motor. The capacitor is located outside the motor housing. The capacitor provides additional starting torque by adjusting the phase angle between the voltage and current in the motor windings. A centrifugal switch is used to disconnect the capacitor and the starting windings from power when the motor reaches a set speed. To reverse the rotation of a capacitor-start motor, the starting winding leads are interchanged. **See Capacitor-Start Motor.**

Dual-Capacitor Motors

Two capacitors give 1ϕ motors additional starting and running torque. One capacitor is sized for starting, and the other capacitor is sized for running. A large-value capacitor is used for starting, and a small-value capacitor is used for running. The starting capacitor is connected in series with the starting winding, and the running capacitor is connected in series with the starting winding after the centrifugal switch opens. To reverse the rotation of a dual-capacitor motor, the starting winding leads are interchanged. **See Dual-Capacitor Motor.**

Chapter 10—Reversing Motor Circuits 125

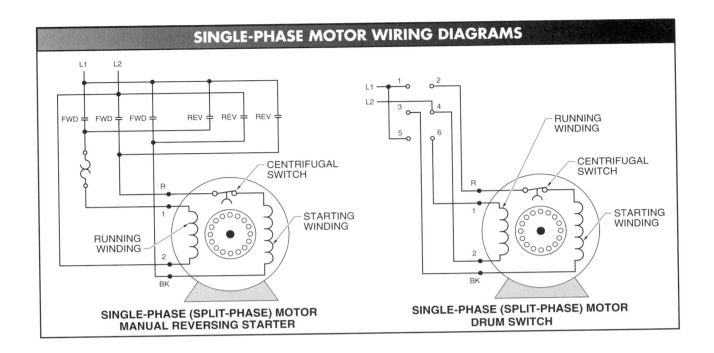

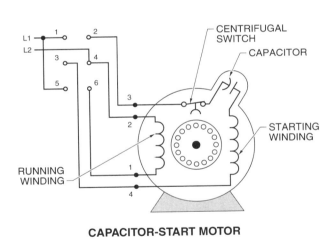

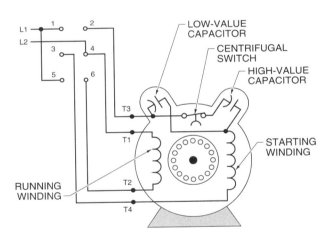

Application 10-4: Reversing Dual-Voltage Motors

Dual-Voltage Motors

Most industrial motors are dual-voltage motors. A *dual-voltage motor* is a motor that is rated at two different voltages. A dual-voltage, 3φ motor is normally rated at 240/480 V. A dual-voltage, 1φ motor is normally rated at 120/240 V. The high voltage is used when possible because the power is the same as at the low voltage but one-half of the current is drawn. With less current drawn, wire size and installation cost are reduced.

To allow a motor to be connected to two different voltages (120/240 V), the main running winding is divided into two parts. The two running windings are connected in series for the high voltage and in parallel for the low voltage. The starting winding is connected across one of the running windings. All motor windings receive nearly the same voltage when wired for low- or high-voltage operation. **See Dual-Voltage Motor.**

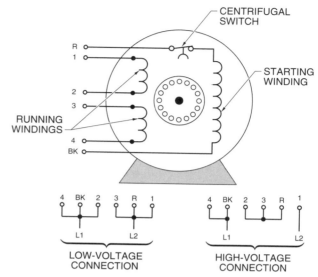

DUAL-VOLTAGE MOTOR

Application 10-5: Reversing DC Motors

DC Motors

The rotation of a DC series, shunt, or compound motor depends on the direction of current flow in the field circuit and the armature circuit. To reverse the rotation of a DC motor, the current direction in the field or in the armature is reversed. Normally, the current in the armature is reversed. **See DC Motor Wiring Diagrams.** Standard abbreviations used with DC motors are A1, A2 = armature; F1, F2 = shunt field; S1, S2 = series field.

A wiring diagram is used to properly wire a DC series motor for reversing. A DC series motor is wired to the starter so that A2 is positive and A1 is negative when the forward contacts are closed, and A2 is negative and A1 is positive when the reverse contacts are closed.

A wiring diagram is used to properly wire a DC shunt motor for reversing. A DC shunt motor is wired to the starter so that A2 is positive and A1 is negative when the forward contacts are closed, and A2 is negative and A1 is positive when the reverse contacts are closed.

A wiring diagram is used for properly wiring a DC compound motor for reversing. A DC compound motor is wired to the starter so that A2 is positive and A1 is negative when the forward contacts are closed, and A2 is negative and A1 is positive when the reverse contacts are closed.

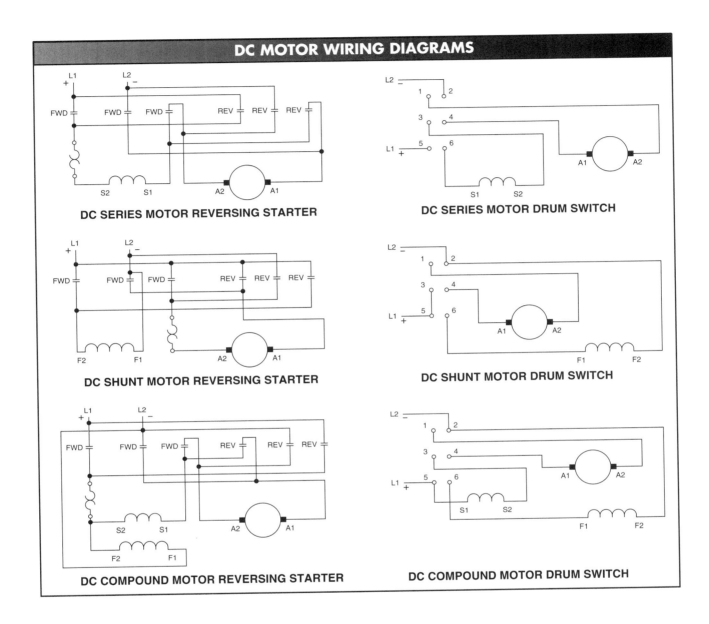

Application 10-6: Troubleshooting Reversing Motor Circuits

Troubleshooting Reversing Circuits

When troubleshooting a reversing circuit, determine whether the motor fails to operate in both directions or only in one direction. Check the overload trip on the motor starter. Reset if tripped and take current measurements on the motor to ensure the motor is not overloaded or the overloads will trip again. Use a DMM set to measure voltage to check the incoming main power and fuses. If there is no power, troubleshoot upstream for a power loss problem and/or change the fuses. If the main power is good, take voltage measurements at the line side (power-in side) of the motor starter. Ensure the voltage is at the correct level. If the voltage is correct, troubleshoot the control circuit. Troubleshoot the control circuit by measuring the incoming voltage to the control circuit. Test the voltage in and out of each component in the control circuit if the incoming voltage to the circuit is correct.

Application 10-7: Hard Wiring Reversing Circuits

Hard Wiring

When installing a motor that requires a reversing circuit, the control devices (pushbuttons, etc.) must be wired into the control circuit. In order to wire the circuit, an electrician uses the line diagram to locate where the wires from the external control devices (pushbuttons, limit switches, etc.) are connected to the control circuit located inside the motor control cabinet.

When using the wiring diagram as a guide in wiring the circuit, a good place to start is at the power source (X1 and X2). First, connect all devices (pushbuttons, etc.) to X1 using the line diagram. Next, connect all devices (overload contacts, etc.) to X2 using the line diagram. Then work through the line diagram moving left to right and top to bottom until all devices are wired. **See Hard Wiring Reversing Circuit.**

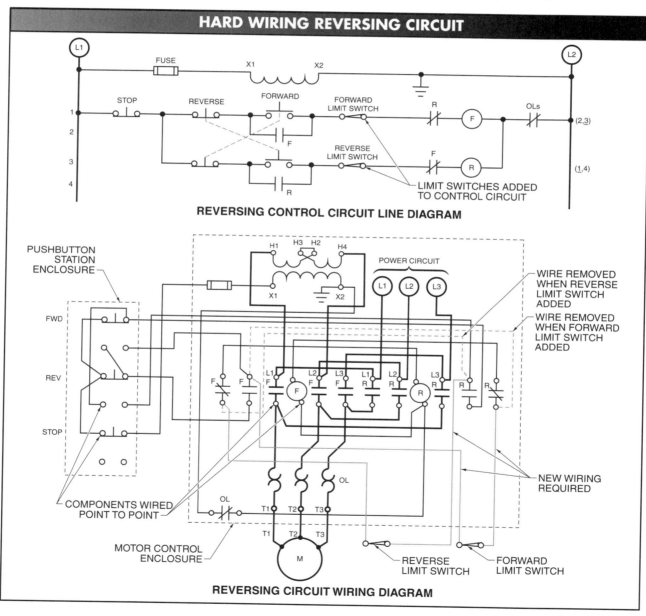

Wiring Circuit Modifications

After a circuit is in operation, modifications are often made in order for the circuit to operate more efficiently for a given application or set of operating conditions. When making circuit modifications, first make the changes that do not require rewiring the circuit. Devices that do not require rewiring include devices that can be directly connected into the circuit without having to open the circuit, such as additional start pushbuttons that can be placed in parallel with existing start pushbuttons. Next, add the devices that require the circuit to be rewired. These devices include stop pushbuttons, overload contacts, and other devices that must be placed in series with existing devices.

Application 10-8: Reversing Circuits and Terminal Strips

Wiring Reversing Circuits to Terminal Strips

Reversing circuits require that the control devices (pushbuttons, etc.) be wired into the control circuit. In order to wire the circuit properly, the line diagram can be used to locate where the wires from the external control devices (pushbuttons, limit switches, etc.) are connected to the control circuit terminal strip located inside the motor control cabinet. The terminal strip numbers follow the wire reference (terminal) numbers used on the line diagram. Some terminals allow more than one wire to be connected at the same terminal location, others use jumpers to connect several terminal points together if several wires need to be connected to the same terminal number. **See Wiring Reversing Circuits Using Terminal Strips.**

Wiring Circuit Modifications to Terminal Strips

After a circuit is in operation, modifications are often made in order for the circuit to operate more efficiently for a given application or set of operating conditions. When using a numbered terminal strip to wire a control circuit, circuit modifications are easily made by using the numbers appearing to the left and right of each device in the line diagram to identify where the device is to be connected at the terminal strip.

Troubleshooting at Terminal Strips

When troubleshooting at terminal strips, approximate meter readings should be anticipated if the meter readings are going to be used to help determine circuit problems. Troubleshooting is started at the control circuit power source (X1 and X2). The terminal numbers that the power source is connected to are identified and voltage measurements taken. If the voltage is not correct, the control circuit fuse(s) and the control circuit transformer (both primary and secondary sides) are checked. If the voltage is correct, one DMM lead is left on the X2 side of the power supply and the other DMM lead is moved through the circuit, moving left to right and top to bottom to determine at which point the power is lost.

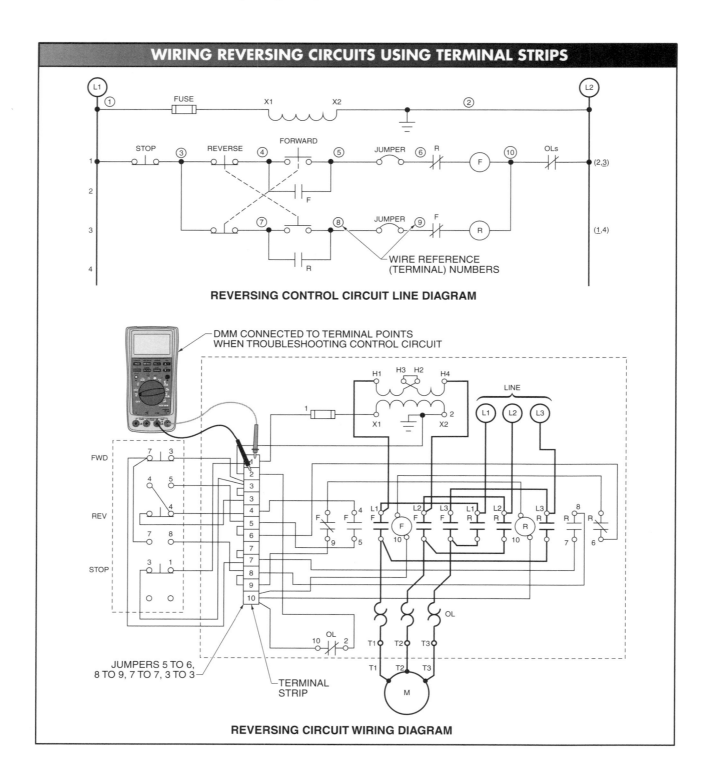

Application 10-9: Reversing Circuits and PLCs

Wiring Reversing Circuit Inputs to PLCs

When installing a motor that requires a reversing circuit using a PLC, the control device inputs (pushbuttons, etc.) must be wired to the PLC input section. In order to wire the circuit, an electrician uses the PLC diagram to locate where the wires from the external control input devices (pushbuttons, limit switches, etc.) connect into the PLC input card. All inputs are identified by an input address (input 1, input 2, etc.) on the PLC input card. Each input is connected to the input address assigned on the PLC programmed line diagram. **See Wiring Reversing Circuits and PLCs.**

Wiring Reversing Circuit Outputs to PLCs

When installing a motor that requires a reversing circuit using a PLC, the control device outputs (motor starter coils, lamps, etc.) must be wired to the PLC output section. In order to wire the circuit, an electrician uses the PLC diagram to locate where the wires from the external output control device (forward starter coil, etc.) connect into the PLC output card. All outputs are identified by an output address (output 1, output 2, etc.) on the PLC output card. Each output is connected to the output address assigned on the PLC programmed line diagram.

132 ELECTRICAL MOTOR CONTROLS *for Integrated Systems* APPLICATIONS MANUAL

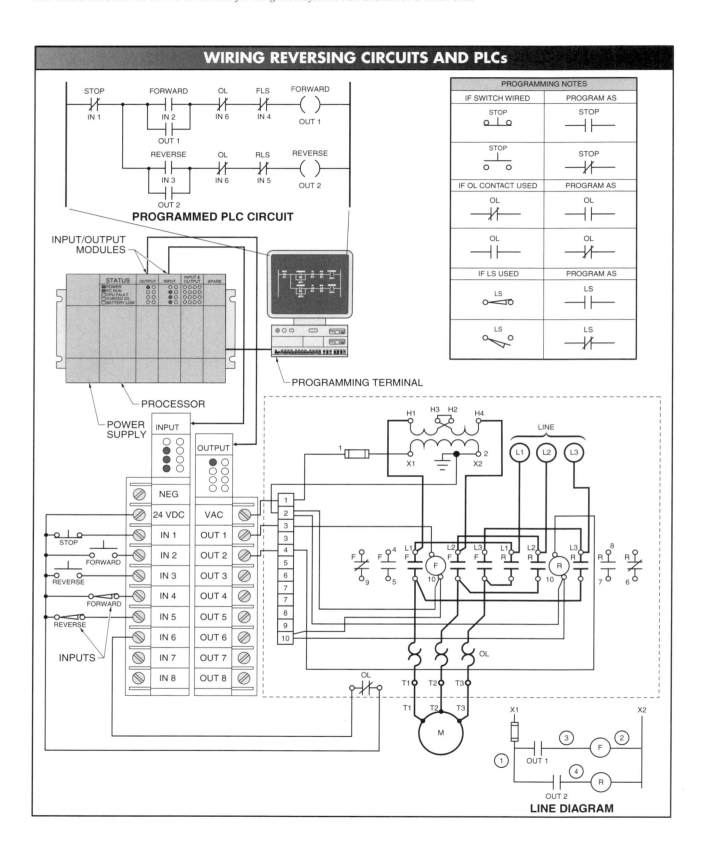

Reversing Motor Circuits

10 Activities

Name _____ Date _____

Activity 10-1: Reversing Motor Circuits

1. Wire the motor for forward and reverse.

ROTATION	L1	L2
FWD	1, 8	4, 5
REV	1, 5	4, 8

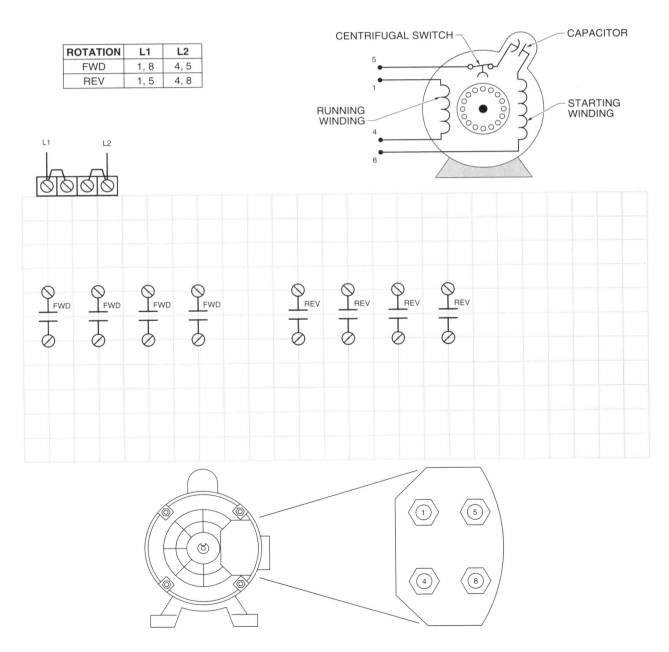

133

Activity 10-2: Reversing Three-Phase Motors

1. Complete the wiring diagram so the motor operates in the forward direction when the forward contacts are closed and in the reverse direction when the reverse contacts are closed.

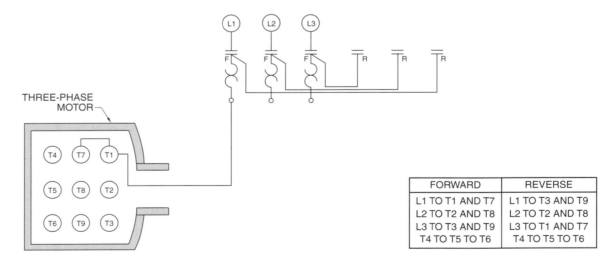

FORWARD	REVERSE
L1 TO T1 AND T7	L1 TO T3 AND T9
L2 TO T2 AND T8	L2 TO T2 AND T8
L3 TO T3 AND T9	L3 TO T1 AND T7
T4 TO T5 TO T6	T4 TO T5 TO T6

Activity 10-3: Reversing Single-Phase Motors

1. Complete the wiring diagram so the motor operates in the forward direction when the forward contacts are closed and in the reverse direction when the reverse contacts are closed.

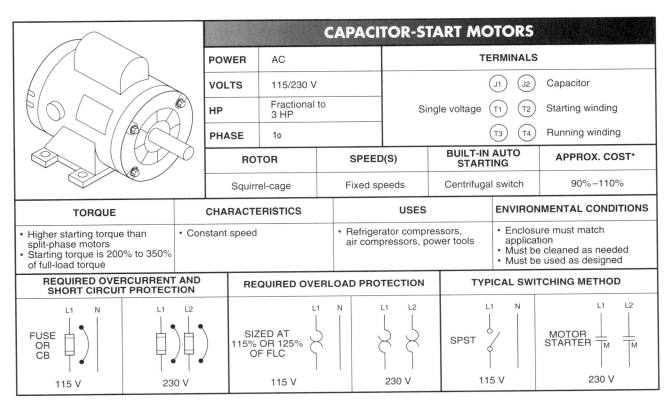

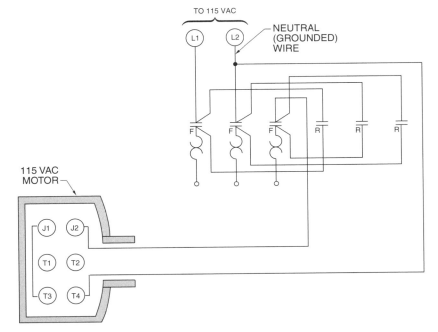

FORWARD	REVERSE
L1 TO T3 AND J1	L1 TO T3 AND J1
L2 TO T4	L2 TO T4
L2 TO T2	L2 TO T1
J2 TO T1	J2 TO T2

Activity 10-4: Reversing Dual-Voltage Motors

1. Wire the motor for forward and reverse operation using low voltage.

ROTATION		L1	L2	JOIN
HIGH VOLTAGE	FWD	1	4, 5	2 & 3 & 8
	REV	1	4, 8	2 & 3 & 5
LOW VOLTAGE	FWD	1, 3, 8	2, 4, 5	—
	REV	1, 3, 5	2, 4, 8	—

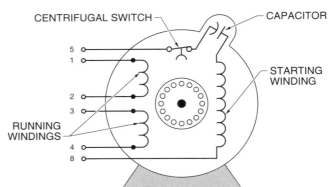

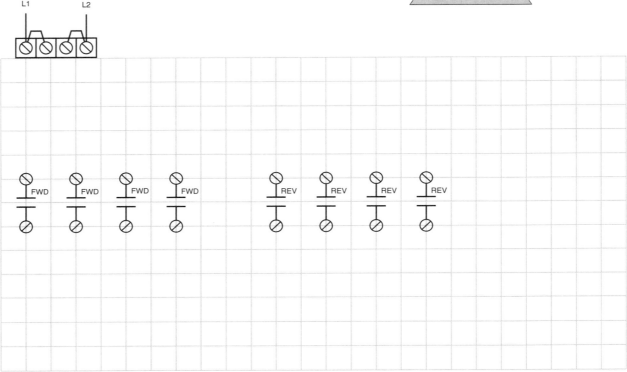

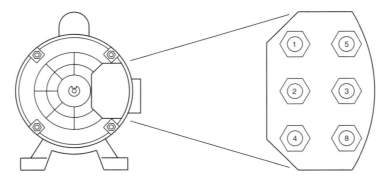

2. Wire the motor for forward and reverse operation using high voltage.

	ROTATION	L1	L2	JOIN
HIGH VOLTAGE	FWD	1	4, 5	2 & 3 & 8
	REV	1	4, 8	2 & 3 & 5
LOW VOLTAGE	FWD	1, 3, 8	2, 4, 5	–
	REV	1, 3, 5	2, 4, 8	–

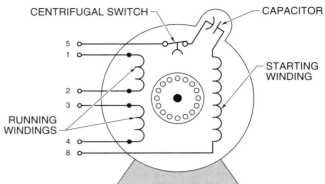
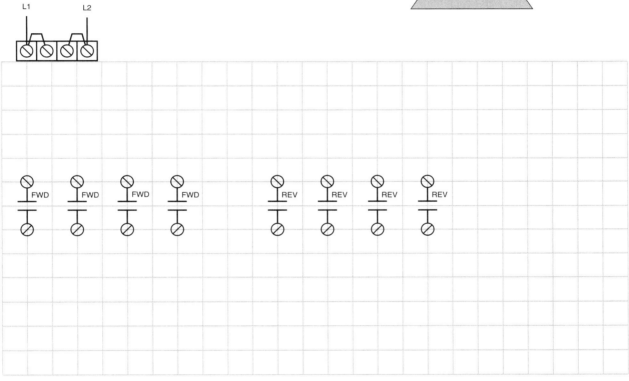
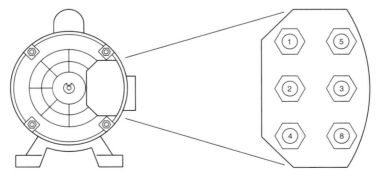

3. Wire the motor for forward and reverse operation using low voltage.

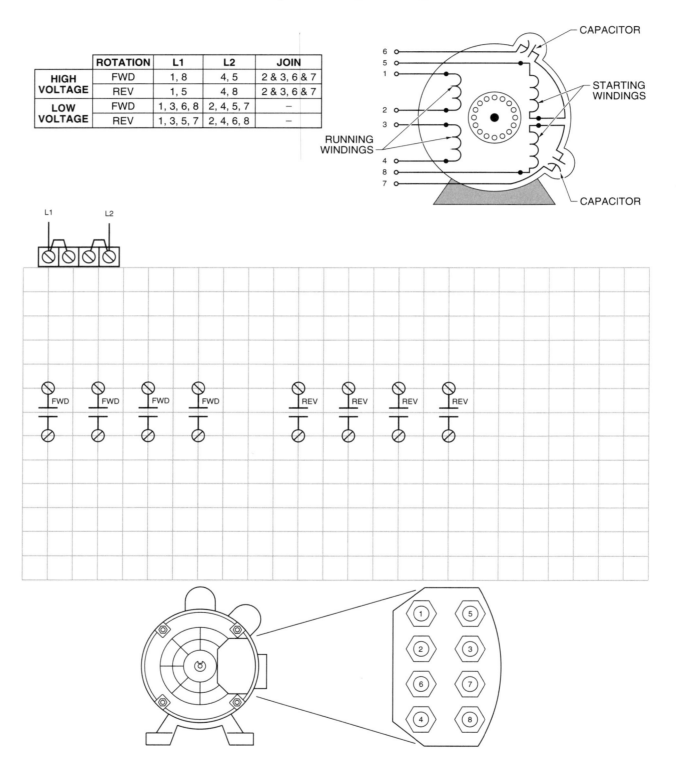

ROTATION		L1	L2	JOIN
HIGH VOLTAGE	FWD	1, 8	4, 5	2 & 3, 6 & 7
	REV	1, 5	4, 8	2 & 3, 6 & 7
LOW VOLTAGE	FWD	1, 3, 6, 8	2, 4, 5, 7	–
	REV	1, 3, 5, 7	2, 4, 6, 8	–

4. Wire the motor for forward and reverse operation using high voltage.

ROTATION		L1	L2	JOIN
HIGH VOLTAGE	FWD	1, 8	4, 5	2 & 3, 6 & 7
	REV	1, 5	4, 8	2 & 3, 6 & 7
LOW VOLTAGE	FWD	1, 3, 6, 8	2, 4, 5, 7	—
	REV	1, 3, 5, 7	2, 4, 6, 8	—

Activity 10-5: Reversing DC Motors

1. Complete the wiring diagram so the motor operates in the forward direction when the forward contacts are closed and in the reverse direction when the reverse contacts are closed.

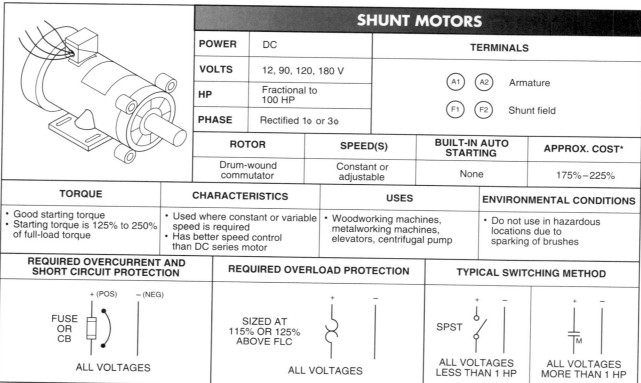

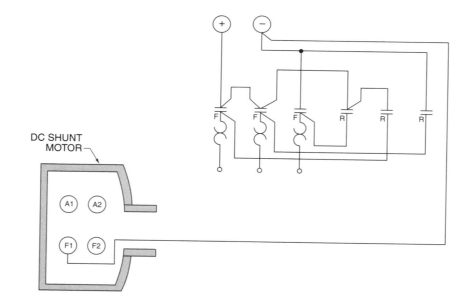

FORWARD	REVERSE
+ TO A2	+ TO A1
− TO A1	− TO A2
− TO F1	− TO F1
+ TO F2	+ TO F2

Activity 10-6: Troubleshooting Reversing Motor Circuits

Determine the malfunctioning component using the Reversing Motor Circuit and the DMM readings. Note: The machine operator reports that the motor operates in the forward direction but not in the reverse direction. The reading on DMM 1 is 36 V when the forward pushbutton is pressed. The reading on DMM 2 is 0 V when the forward or reverse pushbutton is pressed. The reading on DMM 3 is 36 V when the reverse pushbutton is pressed. The reading on DMM 4 is 0 V when the reverse pushbutton is pressed.

_____ **1.** The problem is located in the ___.

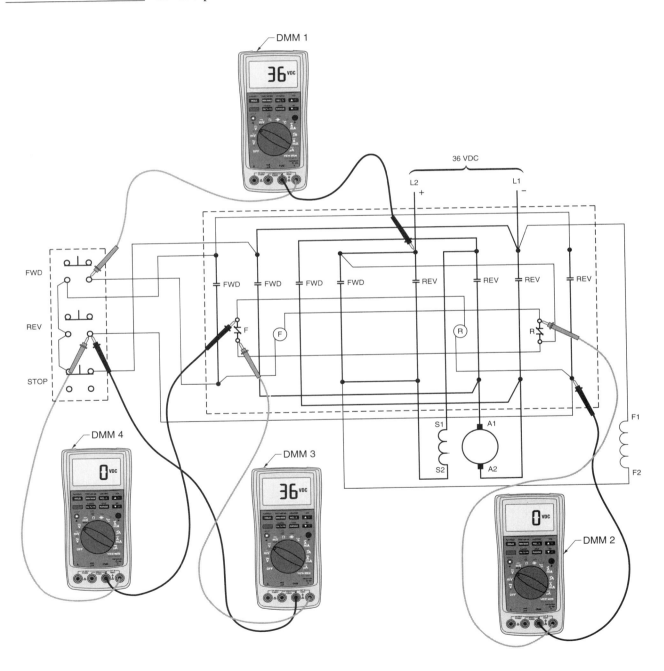

Activity 10-7: Hard Wiring Reversing Circuits

1. Connect the control devices in the pushbutton station into the control circuit inside the motor control enclosure using the line diagram. Use only the minimum number of wires required from the control devices to the enclosure. *Note:* All wires must connect to terminal connection points on the pushbuttons, motor starter, transformer, etc. If two wires need to be connected to the same terminal connection point, place both wires at the terminal connection point.

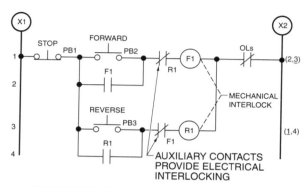

REVERSING CONTROL CIRCUIT LINE DIAGRAM

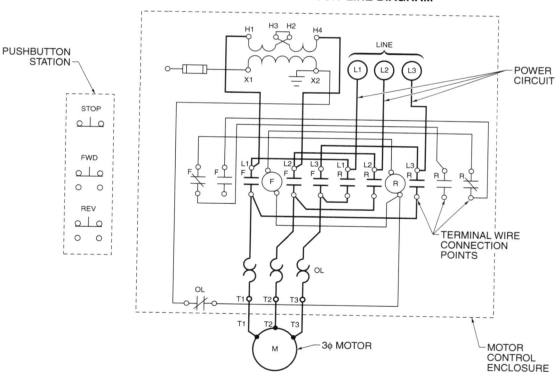

REVERSING CIRCUIT WIRING DIAGRAM

2. Connect the forward (amber) lamp, reverse (red) lamp, and circuit power ON (green) lamp into the control circuit for the reversing motor control circuit using the line diagram. *Note:* All wires must connect to terminal connection points. If two wires need to be connected to the same terminal connection point, place both wires at the terminal connection point.

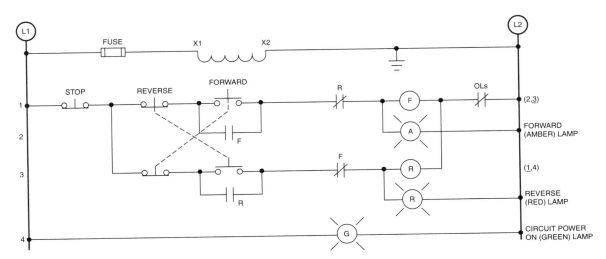

REVERSING CONTROL CIRCUIT LINE DIAGRAM

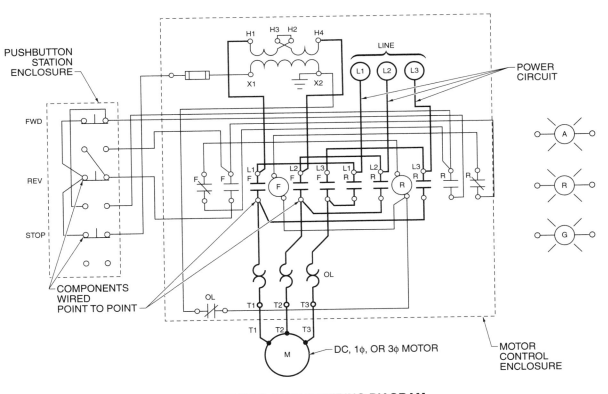

REVERSING CIRCUIT WIRING DIAGRAM

Activity 10-8: Reversing Circuits and Terminal Strips

1. Connect the control devices in the pushbutton station into the control circuit terminal strips located inside the motor control enclosure using the line diagram. Use only the minimum number of wires required from the control devices to the enclosure.

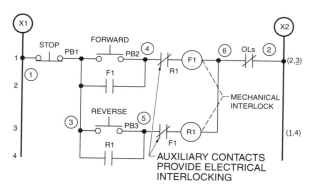

REVERSING CONTROL CIRCUIT LINE DIAGRAM

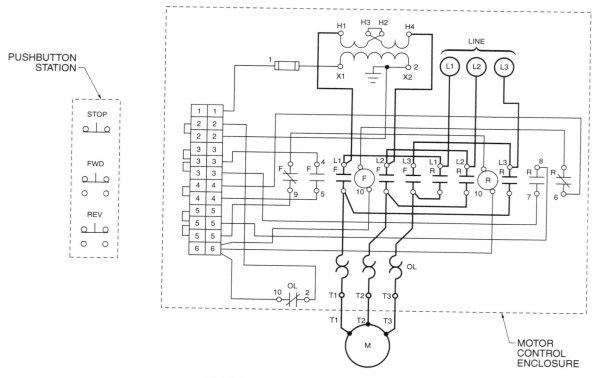

REVERSING CIRCUIT WIRING DIAGRAM

2. Connect the forward (red) lamp, reverse (yellow) lamp, second forward pushbutton, and second reverse pushbutton to the terminal strip located inside the motor control enclosure.

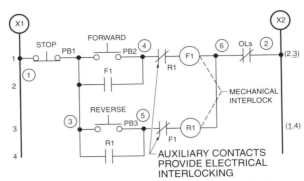

REVERSING CONTROL CIRCUIT LINE DIAGRAM

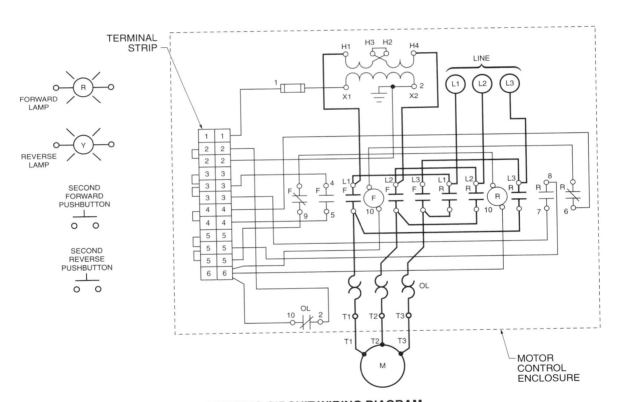

REVERSING CIRCUIT WIRING DIAGRAM

Determine the DMM expected readings if the circuit is operating properly.

_____ 3. The expected DMM reading with the motor OFF is ___ VAC.

_____ 4. The expected DMM reading with the motor running in forward is ___ VAC.

_____ 5. The expected DMM reading with the motor running in reverse is ___ VAC.

_____ 6. The expected DMM reading with the forward and reverse buttons pressed at the same time is ___ VAC.

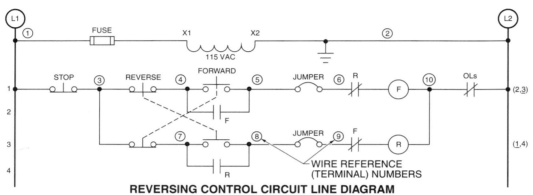
REVERSING CONTROL CIRCUIT LINE DIAGRAM

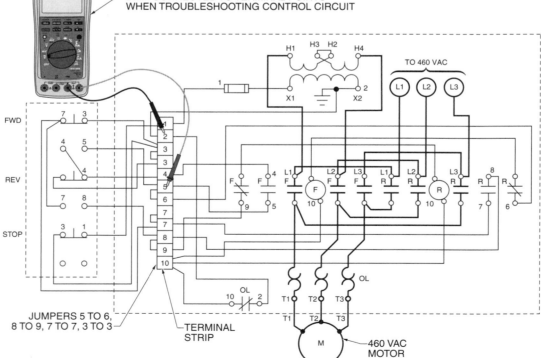
REVERSING CIRCUIT WIRING DIAGRAM

Activity 10-9: Reversing Circuits and PLCs

1. Connect the control devices to the proper PLC input terminals using the PLC diagram. Use only the minimum number of wires required from the control devices to the PLC.

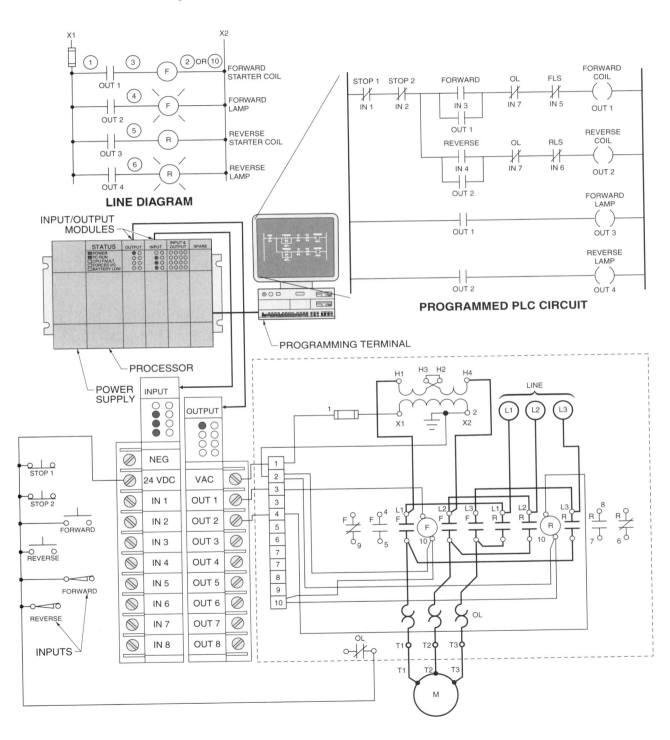

148 ELECTRICAL MOTOR CONTROLS *for Integrated Systems* APPLICATIONS MANUAL

2. Connect the output devices to the proper PLC output terminals using the PLC diagram. Use only the minimum number of wires required from the output devices to the PLC.

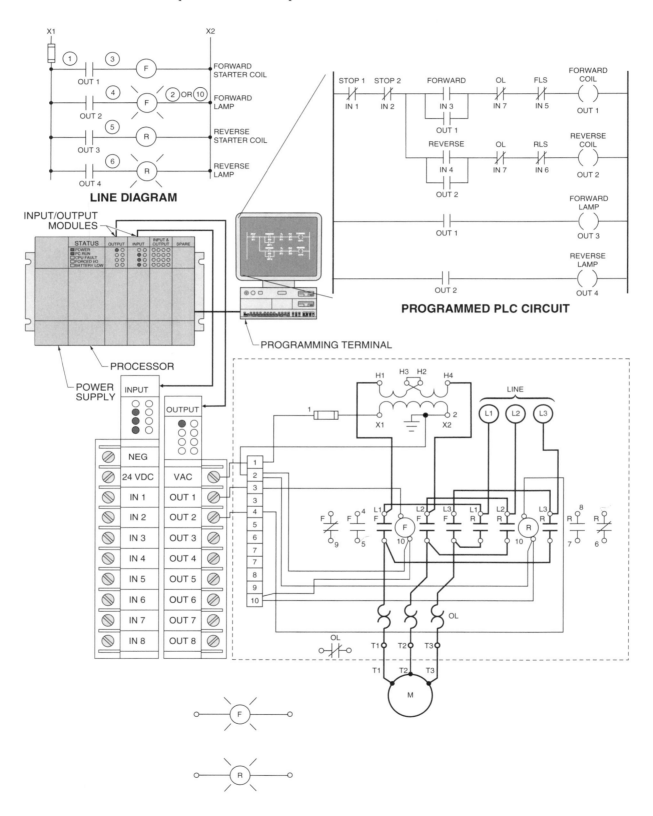

Power Distribution Systems

Applications 11

Application 11-1: Wye and Delta Transformer Configurations

Transformer Configurations

Transformers are connected in wye and delta configurations. A *wye configuration* is a transformer connection that has one end of each transformer coil connected together. The remaining end of each coil is connected to the incoming power lines (primary side) or used to supply power to the load or loads (secondary side). A *delta configuration* is a transformer connection that has each transformer coil connected end-to-end to form a closed loop. Each connection point is connected to the incoming power lines or used to supply power to the load or loads. The voltage output and type available for the load or loads is determined by whether a transformer is connected in a wye or delta configuration. **See Wye and Delta Transformer Configurations.**

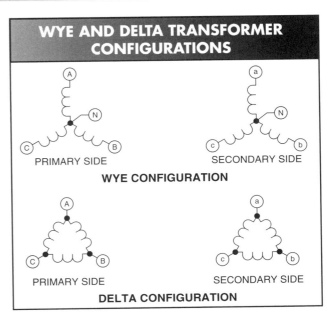

Application 11-2: Motor Control Centers

Motor Control Center Assembly

Applications that require several motor starters assembled in a group normally have the motor starters, disconnect switches, control circuit, and components combined in a motor control center. A *motor control center* is a sheet metal enclosure that encloses and protects motor starters, fuses, circuit breakers, overloads, and wiring. To assemble a motor control center, apply the following procedure:

1. Select the installation location. Locate a motor control center as close to the application and loads as possible because the center is used when troubleshooting the system. When selecting a location, make sure the floor is level and the wall or supporting structure is plumb. **See Installation Location.**

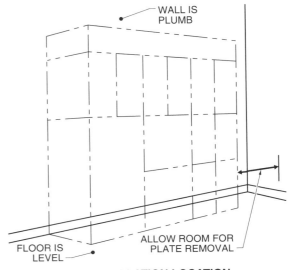

Caution: If the control center is not freestanding, conduit should not be used as the only support for the control center.

2. Assemble the frame. Standard motor control centers consist of units that are 20″ wide by 15″ or 20″ deep and 90″ high. The units are assembled and then secured together.

The control center frame consists of vertical and horizontal supports. The vertical supports are assembled first. The mounting bolts are loosely tightened until all frame supports are in place, then tightened as recommended by the manufacturer. Back or side panels are left off to allow for wiring the unit. **See Frame Assembly.**

3. Install the power bus. Power is distributed through a motor control center by the power bus. The power bus consists of a horizontal bus (main) and vertical buses (units). The horizontal bus runs the length of the center, passing through each unit. Each unit has a vertical bus that is connected to the horizontal bus. The vertical bus runs the full height of each unit. The vertical bus supplies power to each motor starter. The horizontal bus is assembled first, and then the vertical buses are assembled. After the vertical buses are in place, bus supports are installed. **See Bus Installation.**

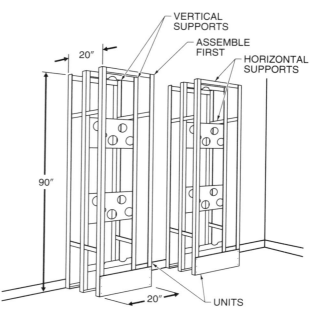

FRAME ASSEMBLY

4. Pull power and control wire. The two types of wiring in a motor control center are power wiring and control wiring. *Power wiring* is any wire connecting the outputs (motors, heating elements, solenoids, lights, etc.). The size of the power wire depends on the amount of current drawn by each load. *Control wiring* is any wire connecting the inputs (pushbuttons, limit switches, pressure switches, etc.). Control wire is normally No. 14 copper wire.

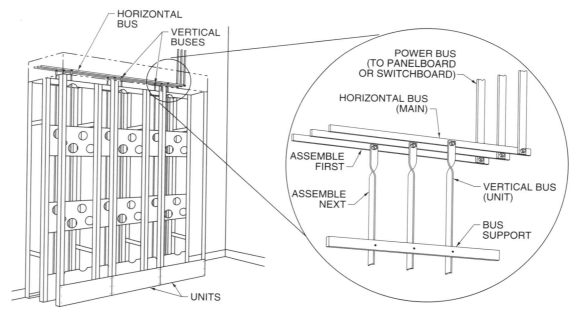

BUS INSTALLATION

The necessary conduit openings are punched and conduit fittings inserted before pulling the wire. After the conduit is installed, enough wire is pulled to reach the most distant circuit for each run. **See Wire Pulling.** *Note:* Wiring must comply with applicable codes and standards.

5. Mount starter modules. Remove necessary knockouts from starter enclosures. Install wire bushings where desired. Mount the largest starter at the lower left, or the unit farthest from the power wiring first. Wire each module one at a time as required.

Individual starter modules are added for each motor. Modules are either fixed or removable. Fixed modules are bolted in. Removable modules are the most common because they are easily inserted and removed.

The most common loads requiring simple and complex control are electric motors. Simplifying and consolidating motor control circuits is required because an electric motor is the backbone of almost all production and industrial applications. To do this, a control center takes the incoming power, control circuitry, required overload and overcurrent protection, and any transformation of power, and combines them into one convenient motor control center.

To meet NEC® standards, each unit must include a fusible disconnect switch or circuit breaker as the branch circuit protective device. A motor control center normally contains only fusible disconnects or only circuit breakers. **See Starter Module Installation.**

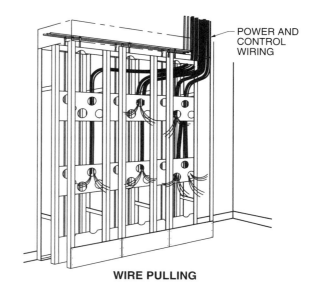

WIRE PULLING

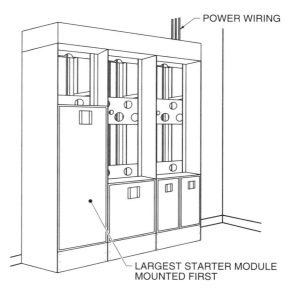

STARTER MODULE INSTALLATION

Application 11-3: Busway Systems

Busway System Support

Supports must be placed with the proper spacing to prevent excessive deflection when installing busway systems. Manufacturer specifications provide data for determining proper spacing of busway hangers. For example, a busway weight of 50 lb with hangers spaced 5′ apart causes 0.70″ of deflection (from Load versus Deflection Specifications). **See Load versus Deflection Specifications.**

Busway systems are composed of elbows, tees, crosses, feeder ducts, and plug-in ducts. The main advantage of busways is that they are easily assembled and disassembled. Each part is designed to bolt together easily and to allow the assembly to be routed in any direction. Elbows and tees allow the assembly to be routed right, left, up, or down. Tees and crosses can also be used to start new sections of the assembly. Changes can be made as the needs of the system change, allowing power to be provided in almost any location in a facility. Per the NEC®, busways (with a few exceptions) shall be installed only in open areas where the assembly is visible. The NEC® also requires busways to be adequately supported. **See Busway System Diagram.**

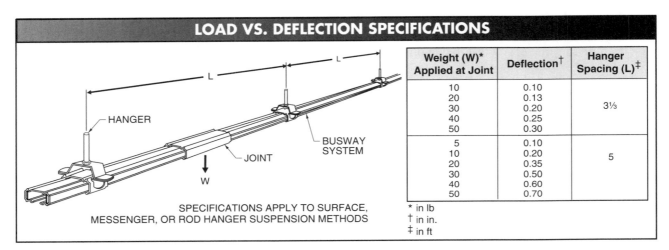

LOAD VS. DEFLECTION SPECIFICATIONS

Weight (W)* Applied at Joint	Deflection†	Hanger Spacing (L)‡
10	0.10	
20	0.13	
30	0.20	3⅓
40	0.25	
50	0.30	
5	0.10	
10	0.20	
20	0.35	5
30	0.50	
40	0.60	
50	0.70	

SPECIFICATIONS APPLY TO SURFACE, MESSENGER, OR ROD HANGER SUSPENSION METHODS

* in lb
† in in.
‡ in ft

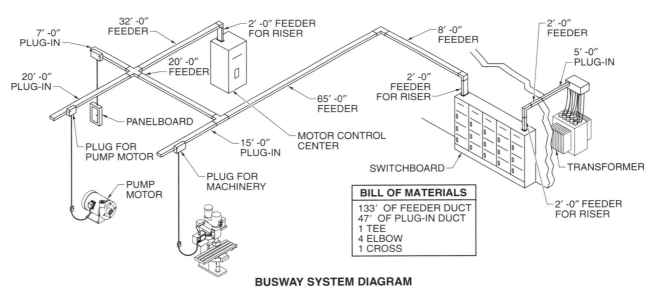

BILL OF MATERIALS
133′ OF FEEDER DUCT
47′ OF PLUG-IN DUCT
1 TEE
4 ELBOW
1 CROSS

BUSWAY SYSTEM DIAGRAM

Application 11-4: Troubleshooting Power Circuits

Power Circuits

A *power circuit* is the part of a circuit that connects the loads to the main power lines. The loads are the devices that convert electrical energy to mechanical energy, heat, light, or sound. Troubleshooting a power circuit is a matter of determining the point in the system at which power is lost. This point may be at the load, at the primary substation, or at any point in between. **See Power Circuit Distribution System.** To troubleshoot a power circuit, apply the following procedure:

1. Check fuses, circuit breakers, and overload contacts. All power circuits are protected from overcurrents at several points in the system by fuses or circuit breakers. For individual load problems, start with the fuses, circuit breakers, and overloads (on motor starters) closest to the load, and then work back through the system. For multiload problems, start with the motor control center, secondary switchboard, lighting panel, or power panel that feeds the loads.

2. Check control circuit. When energizing small loads, the control circuit may directly switch the power to the load. When energizing large loads, the control circuit uses an interface such as a contactor or motor starter to energize the load. Check to ensure the control circuit is delivering power to the load. If the control circuit is delivering power as required, the problem is not in the control circuit. If the control circuit is not delivering power as required, the problem is in the control circuit.

3. Check load. If power is delivered to the load but the load does not work, the problem is in the load. If the power delivered to the load is correct, the load needs replacement or service.

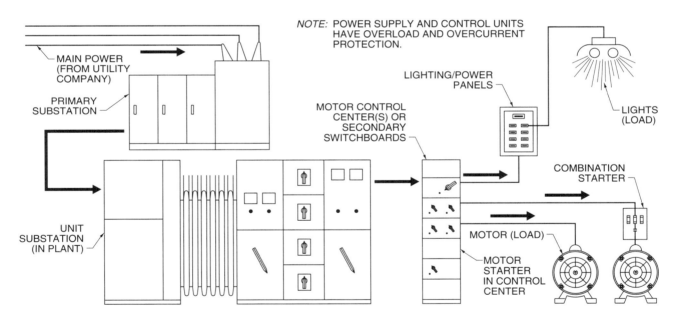

POWER CIRCUIT DISTRIBUTION SYSTEM

Application 11-5: 120/240 V, Single-Phase Systems

120/240 V, Single-Phase, 3-Wire Systems

A 120/240 V, 1ϕ, 3-wire system is used to supply power to users that require 120 V and 240 V, 1ϕ power. This system provides 120 V, 1ϕ; 240 V, 1ϕ; and 120/240 V, 1ϕ circuits. The neutral wire is grounded and should not be fused or switched at any point.

A 120/240 V, 1ϕ, 3-wire system is commonly used for interior wiring for lighting and small appliance use. For this reason, a 120/240 V, 1ϕ, 3-wire system is the primary system used to supply most residential buildings. In commercial applications, small appliance usage is greater than in a typical residential application, making this system ideal for use with small commercial applications, even though a larger power panel or additional (multiple) panels are used. **See 120/240 V, Single-Phase Systems.**

In the past, a 120/240 V, 1ϕ, 3-wire system was used to supply power to 1ϕ loads such as lamps, motors, televisions, and computers. Modern electric motor drives are available that can connect to a 120/240 V, 1ϕ circuit and deliver 3ϕ power to a 3ϕ motor. Three-phase motors can be used in residential applications. For example, residential HVAC systems use 3ϕ motors to save energy cost, even if only 1ϕ power is available. Electric motor drives that are rated for a 1ϕ input (with a 3ϕ output) are generally limited to 3ϕ motors of 5 HP or less. A 3ϕ, 5 HP motor provides enough power for most industrial and commercial motor drive applications.

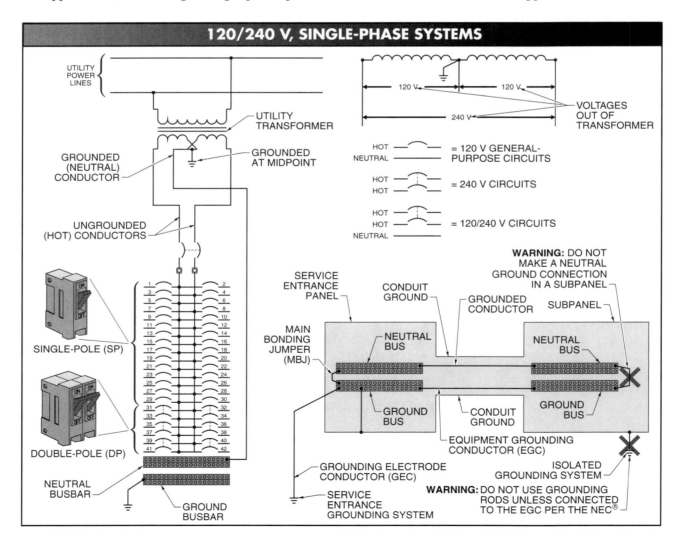

Application 11-6: 120/208 V, Three-Phase Systems

120/208 V, Three-Phase, 4-Wire Systems

A 120/208 V, 3ϕ, 4-wire system is used to supply users that require a large amount of 120 V, 1ϕ power and some low-voltage, 3ϕ power. This system includes three ungrounded (hot) lines and one grounded (neutral) line. Each hot line has 120 V to ground when connected to the neutral line. The 120 V circuits should be balanced to distribute the power equally among the three hot lines. This is accomplished by alternately connecting the 120 V circuits to the power panel so each phase (A to N, B to N, and C to N) is divided among the loads (lamps, receptacles, etc.). The 208 V, 1ϕ loads such as heating elements and motor drives should be balanced between phases (A to B, B to C, and C to A). **See 120/208 V, Three-Phase Systems.**

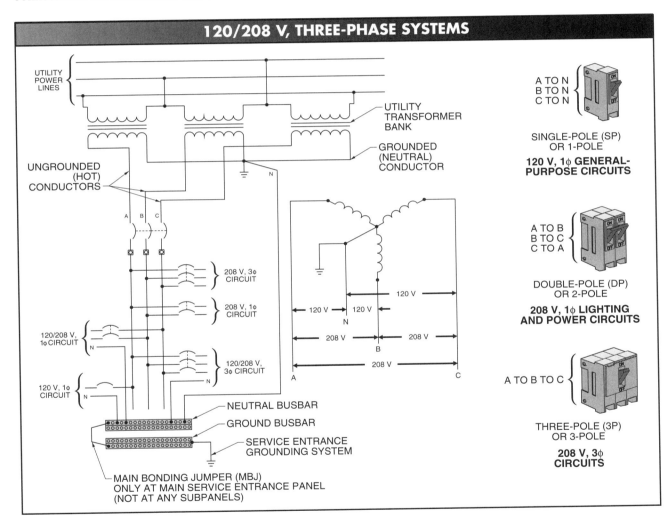

Application 11-7: 120/240 V, Three-Phase Systems

120/240 V, Three-Phase, 4-Wire Systems

A 120/240 V, 3ϕ, 4-wire system is used to supply users that require large amounts of 3ϕ power with some 120 V and 240 V, 1ϕ power. This system supplies 1ϕ power delivered by one of the three transformers and 3ϕ power delivered by using all three transformers. The 1ϕ power is provided by center-tapping one of the transformers. Because only one transformer delivers all of the 1ϕ power, this system is used in applications that require mostly 3ϕ power or 240 V, 1ϕ power and some 120 V, 1ϕ power. This system works because, in many commercial applications, the total amount of 1ϕ power used is small when compared to the total amount of 3ϕ power used. Each transformer may be center-tapped if a large amount of 1ϕ power is required. Also available is approximately 195 V, 1ϕ power from B to N. This voltage is considered hazardous (too high) because it could damage equipment and cause injury to personnel. The voltage is too high for low-voltage, 1ϕ loads (115/120 VAC) and too low for high-voltage, 1ϕ loads (230/240 VAC). Per NEC® 215.8, the higher voltage phase (B) should be colored orange. By marking the higher voltage phase, the B to N voltage can be recognized as hazardous, helping to avoid equipment and safety problems. **See 120/240 V, Three-Phase Systems.**

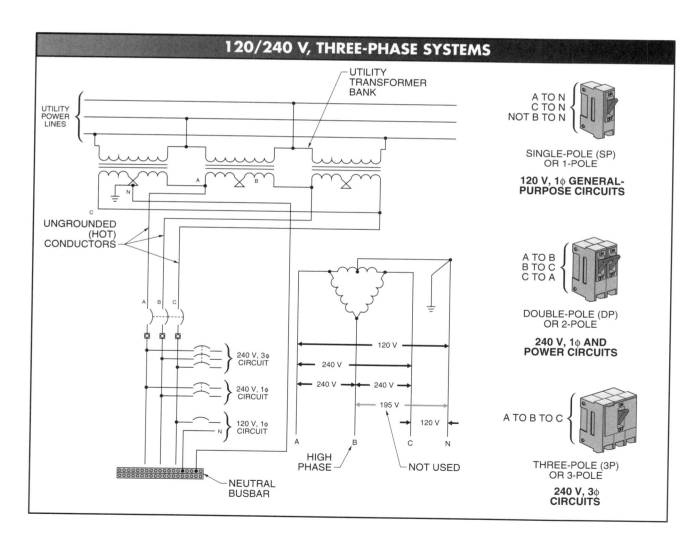

Application 11-8: 277/480 V, Three-Phase Systems

277/480 V, Three-Phase, 4-Wire Systems

A 277/480 V, 3φ, 4-wire system is the same as a 120/240 V, 3φ, 4-wire system with the exception of higher voltage levels. This system includes three ungrounded (hot) lines and one grounded (neutral) line. Each hot line has 277 V to ground when connected to the neutral or 480 V when connected between any two hot (A to B, B to C, or C to A) lines.

A 277/480 V, 3φ, 4-wire system provides 277 V, 1φ or 480 V, 1φ power. For this reason, a 277/480 V, 3φ, 4-wire system is not used to supply 120 V, 1φ general lighting and appliance circuits. A 277/480 V, 3φ, 4-wire system can be used to supply 277 V and 480 V, 1φ lighting circuits. Such high-voltage lighting circuits are used in commercial fluorescent and high-intensity-discharge (HID) lighting circuits.

A system that cannot deliver 120 V, 1φ power appears to have limited use. In many commercial applications (recreational complexes, school campuses, offices, parking lots, etc.), lighting is a major part of the electrical system. Because large commercial applications include several sets of transformer banks, 120 V, 1φ power is available through other transformers. Additional transformers can also be connected to the 277/480 V, 3φ, 4-wire system to reduce the voltage to 120 V, 1φ.

When motor drives are used for large commercial HVAC systems, the drive is connected to the 480 V, 3φ power lines. A motor drive should be connected to a 3φ power supply, even if the drive can be connected to a 1φ supply. This ensures that the internal parts of the drive are evenly balanced to carry only one-third of the total power per line. **See 277/480 V, Three-Phase Systems.**

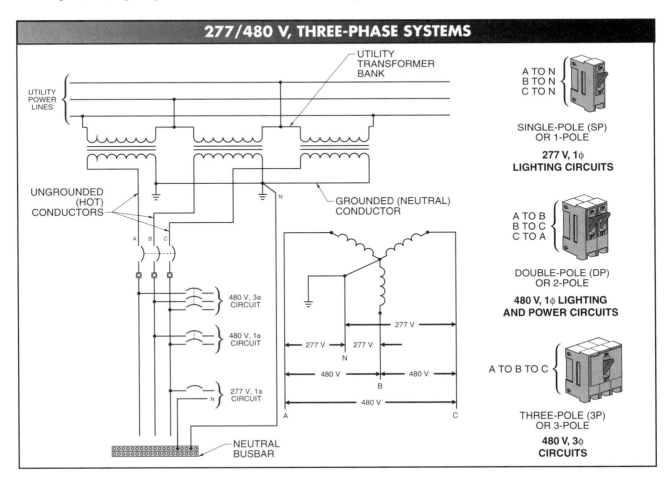

Application 11-9: Plug and Receptacle Configurations and Ratings

Plugs and Receptacles

A *plug* is a device connected to the end of a cord that includes male contacts, which are inserted into a receptacle in order to establish an electrical connection. A *receptacle (outlet)* is a device with female contacts, which are used to connect a plug to the hardwired electrical distribution system.

Plugs and receptacles are marked with ratings for maximum voltage and amperes (125 V/15 A, 125 V/20 A, 250 V/20 A, etc.). The listed rating indicates the maximum voltage and current that can be applied to the plug or receptacle. For example, a 125 V, 15 A rated plug or receptacle can be used on a 115 V, 120 V, etc. system that carries 0 A to 15 A.

The rating of a plug or receptacle is also indicated by the physical configuration of the plug or receptacle. Plug and receptacle configurations ensure that an electrical load connected with a plug is connected to a receptacle that is correct for the load's voltage and current ratings. The configuration is designed to prevent a higher current or different voltage plug from being connected to a lesser (or incorrect) voltage receptacle.

Plugs and receptacles follow standard configurations. The National Electrical Manufacturers Association (NEMA) has established a set of standard plug and receptacle configurations that identify the type and rating of terminations. A proper receptacle has a high enough voltage and current rating to safely deliver enough power to the load. **See Nonlocking Wiring Devices.**

NON-LOCKING WIRING DEVICES

2-POLE, 3-WIRE

Wiring Diagram	NEMA ANSI	Receptacle Configuration	Rating
	5-15 C73.11		15 A 125 V
	5-20 C73.12		20 A 125 V
	5-30 C73.45		30 A 125 V
	5-50 C73.46		50 A 125 V
	6-15 C73.20		15 A 250 V
	6-20 C73.51		20 A 250 V
	6-30 C73.52		30 A 250 V
	6-50 C73.53		50 A 250 V
	7-15 C73.28		15 A 277 V
	7-20 C73.63		20 A 277 V
	7-30 C73.64		30 A 277 V
	7-50 C73.65		50 A 277 V

Power Distribution Systems

Activity 11-1: Wye and Delta Transformer Configurations

Answer questions 1–7 using Transformer Bank 1.

_____ 1. The primary side of the transformers is connected in a(n) ___ configuration.

_____ 2. The secondary side of the transformers is connected in a(n) ___ configuration.

List the power lines to which the loads are connected. If a load cannot be correctly connected to the transformer bank, mark it not possible.

_____ 3. A 120 VAC motor is connected to ___, ___, or ___.

_____ 4. A 230 VAC (208 VAC to 240 VAC), 1φ motor is connected to ___, ___, or ___.

_____ 5. A dual-voltage 115/230 VAC, 1φ motor is connected to ___, ___, or ___ for 115 V.

_____ 6. A dual-voltage 230/460 VAC, 3φ motor is connected to ___, ___, or ___ for 230 V.

_____ 7. A dual-voltage 230/460 VAC, 3φ motor is connected to ___, ___, or ___ for 460 V.

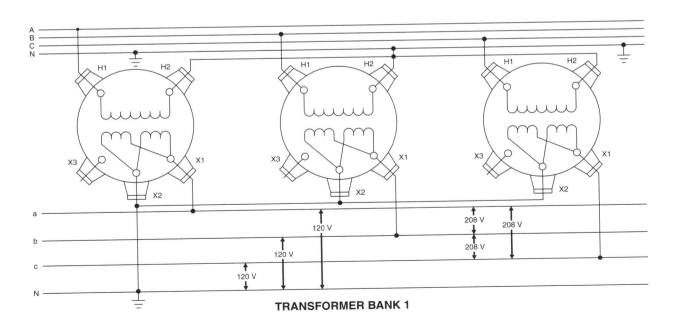

TRANSFORMER BANK 1

Answer questions 8–14 using Transformer Bank 2.

_____ **8.** The primary side of the transformers is connected in a(n) ___ configuration.

_____ **9.** The secondary side of the transformers is connected in a(n) ___ configuration.

List the power lines to which the loads are connected. If a load cannot be correctly connected to the transformer bank, mark it not possible.

_____ **10.** A 120 VAC motor is connected to ___, ___, or ___.

_____ **11.** A 230 VAC (208 VAC to 240 VAC), 1ϕ motor is connected to ___, ___, or ___.

_____ **12.** A dual-voltage 115/230 VAC motor is connected to ___, ___, or ___ for 115 V.

_____ **13.** A dual-voltage 115/230 VAC motor is connected to ___, ___, or ___ for 230 V.

_____ **14.** A dual-voltage 230/460 VAC, 3ϕ motor is connected to ___, ___, or ___ for 460 V.

TRANSFORMER BANK 2

Answer questions 15–21 using Transformer Bank 3.

_____ **15.** The primary side of the transformers is connected in a(n) ___ configuration.

_____ **16.** The secondary side of the transformers is connected in a(n) ___ configuration.

List the power lines to which the loads are connected. If a load cannot be correctly connected to the transformer bank, mark it not possible.

_____ **17.** A 120 VAC motor is connected to ___ or ___.

_____ **18.** A 240 VAC (208 VAC to 240 VAC), 1ϕ motor is connected to ___, ___, or ___.

_____ **19.** A dual-voltage 120/240 VAC motor is connected to ___ or ___ for 120 V.

_____ **20.** A dual-voltage 120/240 VAC, 3ϕ motor is connected to ___, ___, or ___ for 240 V.

_____ **21.** A dual-voltage 240/480 VAC, 3ϕ motor is connected to ___, ___, or ___ for 480 V.

TRANSFORMER BANK 3

Answer questions 22–29 using Transformer Bank 4.

_____ **22.** The primary side of the transformers is connected in a(n) ___ configuration.

_____ **23.** The secondary side of the transformers is connected in a(n) ___ configuration.

List the power lines to which the loads are connected. If a load cannot be correctly connected to the transformer bank, mark it not possible.

_____ **24.** A 115 VAC motor is connected to ___.

_____ **25.** A 230 VAC (208 VAC to 240 VAC), 1ϕ motor is connected to ___ or ___.

_____ **26.** A dual-voltage 115/230 VAC motor is connected to ___ or ___ for 115 V.

_____ **27.** A dual-voltage 115/230 VAC motor is connected to ___ or ___ for 230 V.

_____ **28.** A dual-voltage 230/460 VAC, 3ϕ motor is connected to ___ or ___ for 230 V.

_____ **29.** A dual-voltage 230/460 VAC, 3ϕ motor is connected to ___, ___, or ___ for 460 V.

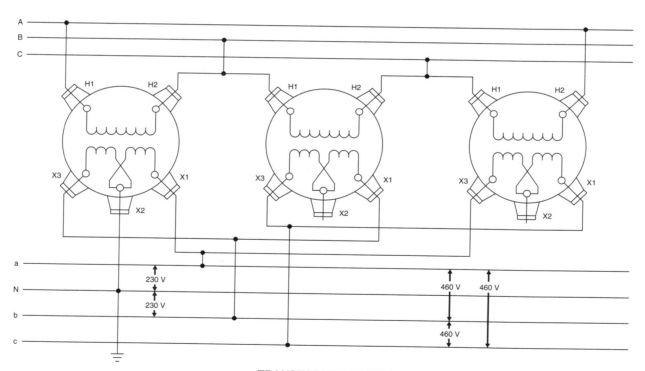

TRANSFORMER BANK 4

Activity 11-2: Motor Control Centers

Answer the questions using the Motor Control Center Assembly on pages 149-151.

1. What should be given consideration when selecting a location for the motor control center?

2. How should the motor control center be secured to prevent movement?

3. How are the busbars of adjacent buses connected?

4. How much wire is required in the motor control center for the preliminary wiring?

5. In what order should the starters be mounted in the control center?

Activity 11-3: Busway Systems

Answer questions 1–17 using the Load versus Deflection Specifications on page 152.

_____ 1. A busway weight of 30 lb with hangers spaced 5′ apart causes ___″ of deflection.

_____ 2. A busway weight of 30 lb with hangers spaced 3.33′ apart causes ___″ of deflection.

_____ 3. The hanger spacing required for a 5 lb busway if a maximum deflection of no more than 0.18″ is allowed is ___′.

_____ 4. The hanger spacing required for a 15 lb busway if a maximum deflection of no more than 0.18″ is allowed is ___′.

_____ 5. The hanger spacing required for a 20 lb busway if a maximum deflection of no more than 0.18″ is allowed is ___′.

_____ 6. The hanger spacing required for a 30 lb busway if a maximum deflection of no more than 0.18″ is allowed is ___′.

164 ELECTRICAL MOTOR CONTROLS *for Integrated Systems* APPLICATIONS MANUAL

_____ 7. The hanger spacing required for a 50 lb busway if a maximum deflection of no more than 0.18″ is allowed is ___′.

_____ 8. The hanger spacing required for a 5 lb busway if a maximum deflection of no more than 0.65″ is allowed is ___′.

_____ 9. The hanger spacing required for a 10 lb busway if a maximum deflection of no more than 0.65″ is allowed is ___′.

_____ 10. The hanger spacing required for a 15 lb busway if a maximum deflection of no more than 0.65″ is allowed is ___′.

_____ 11. The hanger spacing required for a 20 lb busway if a maximum deflection of no more than 0.65″ is allowed is ___′.

_____ 12. The hanger spacing required for a 40 lb busway if a maximum deflection of no more than 0.65″ is allowed is ___′.

_____ 13. The hanger spacing required for a 50 lb busway if a maximum deflection of no more than 0.65″ is allowed is ___′.

_____ 14. A maximum of ___ lb/ft of a busway can be connected to the hangers if the busway system is supported every 5′, and no more than 0.70″ of deflection is allowed.

15. Complete the bill of materials for the busway system using the pricing information and dimensions.

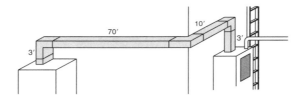

PRICING INFORMATION (LIST)	
Feeder busway duct	$40.00 per ft
Plug-in busway duct	45.00 per ft
Elbows	170.00 ea
Tees	190.00 ea
Crosses	225.00 ea
50 A breakers	175.00 ea
100 A breakers	225.00 ea

BILL OF MATERIALS

Totals

_____ ft of feeder duct @ $_____ per ft   _____

_____ ft of plug-in duct @ $_____ per ft   _____

_____ elbows @ $_____ ea   _____

_____ tees @ $_____ ea   _____

_____ crosses @ $_____ ea   _____

_____ 50 A breakers @ $_____ ea   _____

_____ 100 A breakers @ $_____ ea   _____

SUBTOTAL _____ (LIST)

−20% DISCOUNT _____

TOTAL _____

16. Complete the bill of materials for the busway system using the pricing information and dimensions.

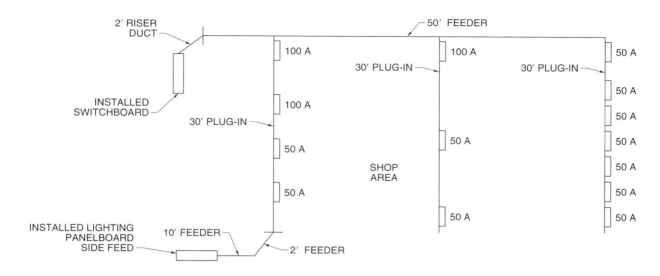

PRICING INFORMATION (LIST)

Feeder busway duct	$40.00 per ft
Plug-in busway duct	45.00 per ft
Elbows	170.00 ea
Tees	190.00 ea
Crosses	225.00 ea
50 A breakers	175.00 ea
100 A breakers	225.00 ea

BILL OF MATERIALS

	Totals
_____ ft of feeder duct @ $_____ per ft	_____
_____ ft of plug-in duct @ $_____ per ft	_____
_____ elbows @ $_____ ea	_____
_____ tees @ $_____ ea	_____
_____ crosses @ $_____ ea	_____
_____ 50 A breakers @ $_____ ea	_____
_____ 100 A breakers @ $_____ ea	_____
SUBTOTAL	_____ (LIST)
−20% DISCOUNT	_____
TOTAL	_____

Activity 11-4: Troubleshooting Power Circuits

For each reported problem, identify the best location in the Power Circuit to begin troubleshooting (checking circuit breaker, taking electrical measurements, etc.) from the following list of circuit locations. Some problems may have more than one location to begin troubleshooting.

_____ 1. It is reported that motor 2 is making a loud, high-pitched noise.

_____ 2. It is reported the loads in the area serviced by panelboard 2 are not working.

_____ 3. It is reported that the lamp loads connected to panelboard 2 are dimming, and that computer systems are crashing.

_____ 4. It is reported that motor 1 cannot turn ON.

_____ 5. It is reported that during the plant's peak power usage, some loads seem to have a low-voltage condition.

A. Main Power from Utility Company (outside switchgear)
B. Panelboard 1 Circuit Breaker in Switchboard
C. Panelboard 2 Circuit Breaker in Switchboard
D. Panelboard 1 Main Circuit Breaker
E. Panelboard 2 Main Circuit Breaker
F. Circuit Fuse for Motor 1 Branch Circuit
G. Motor Control 1
H. Motor 1
I. Circuit Fuse for Motor 2 Branch Circuit
J. Motor Control 2
K. Motor 2
L. Circuit Breaker for Circuit 1 Load
M. Circuit Breaker for Circuit 2 Load
N. Circuit Breaker for Circuit 3 Load
O. Circuit Breaker for Circuit 4 Load
P. Circuit 1 Load
Q. Circuit 2 load
R. Circuit 3 Load
S. Circuit 4 Load
T. Transformer 1
U. Transformer 2
V. Transformer 3

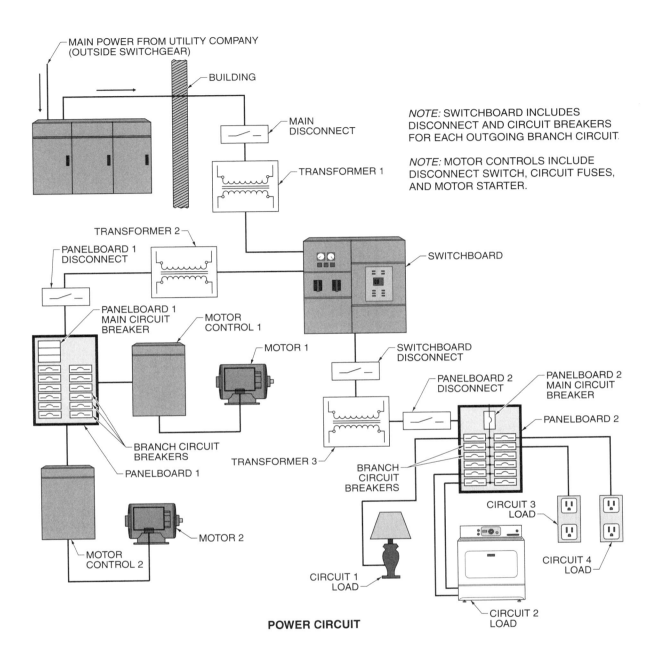

POWER CIRCUIT

Activity 11-5: 120/240 V, Single-Phase Systems

When troubleshooting 120/240 V, 1ϕ systems, approximate meter readings should be anticipated if the meter readings are going to be used to help determine circuit problems. Determine the expected DMM readings if the circuit is operating properly using 120/240 V, Single-Phase Systems on page 154.

_____ 1. The expected reading of DMM 1 is ___ VAC.

_____ 2. The expected reading of DMM 2 is ___ VAC.

_____ 3. The expected reading of DMM 3 is ___ VAC.

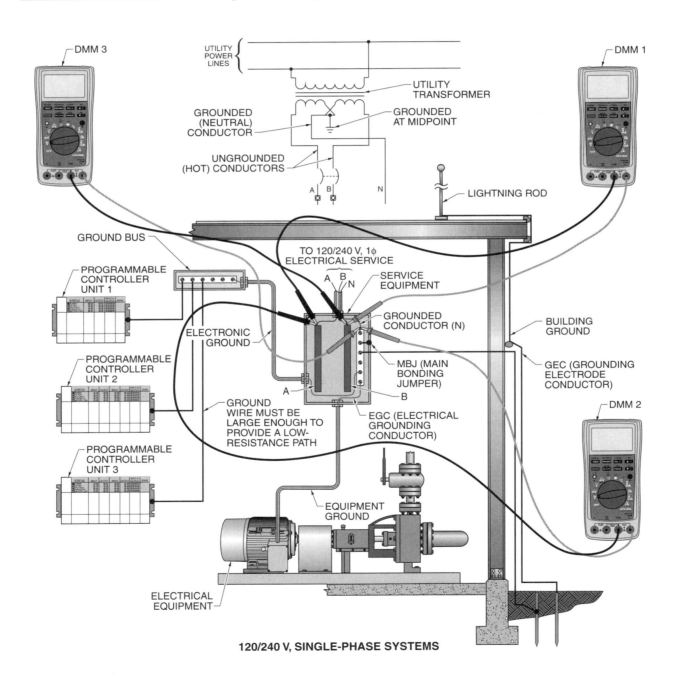

120/240 V, SINGLE-PHASE SYSTEMS

Activity 11-6: 120/208 V, Three-Phase Systems

When troubleshooting 120/208 V, 3ϕ systems, approximate meter readings should be anticipated if the meter readings are going to be used to help determine circuit problems. Determine the expected DMM readings if the circuit is operating properly using 120/208 V, Three-Phase Systems on page 155.

_____ 1. The expected reading of DMM 1 is ___ VAC.

_____ 2. The expected reading of DMM 2 is ___ VAC.

_____ 3. The expected reading of DMM 3 is ___ VAC.

_____ 4. The expected reading of DMM 4 is ___ VAC.

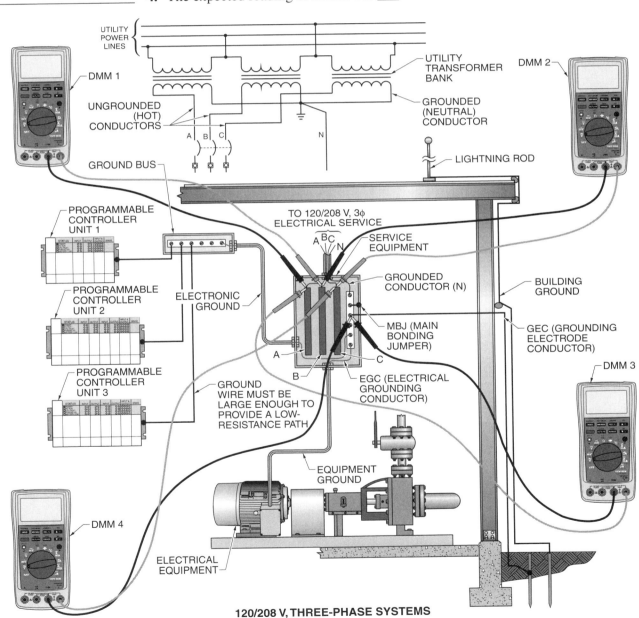

120/208 V, THREE-PHASE SYSTEMS

Activity 11-7: 120/240 V, Three-Phase Systems

When troubleshooting 120/240 V, 3φ systems, approximate meter readings should be anticipated if the meter readings are going to be used to help determine circuit problems. Determine the expected DMM readings if the circuit is operating properly using *120/240 V, Three-Phase Systems* on page 156.

_____ 1. The expected reading of DMM 1 is ___ VAC.

_____ 2. The expected reading of DMM 2 is ___ VAC.

_____ 3. The expected reading of DMM 3 is ___ VAC.

_____ 4. The expected reading of DMM 4 is ___ VAC.

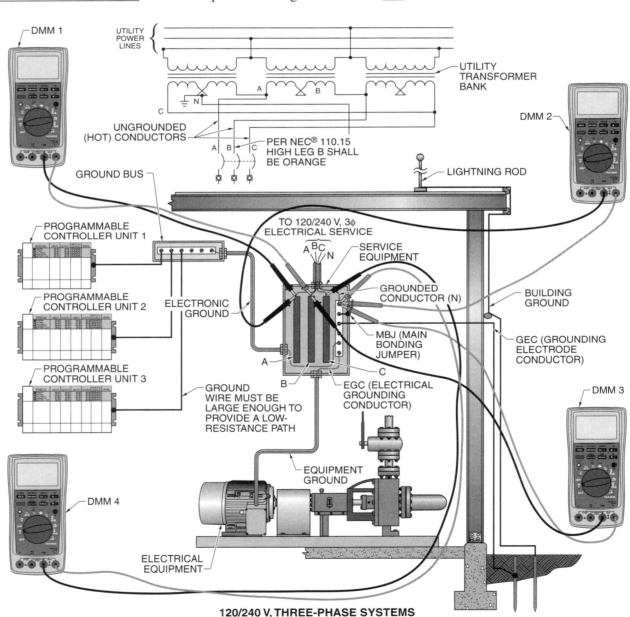

120/240 V, THREE-PHASE SYSTEMS

Activity 11-8: 277/480 V, Three-Phase Systems

Electrical loads (motors, electric heaters, lamps, etc.) are ordered from the manufacturer so that they can be used on different voltage sources (120 V, 208 V, 240 V, etc.). For example, high-intensity-discharge (HID) lamps are available in many voltage types to cover different residential, commercial, and industrial lighting requirements. Answer the following questions using the HID Lamp Ordering Information.

_____ 1. For the given service, can a model number HID-01 lamp be connected to the service panel voltage?

_____ 2. For the given service, can a model number HID-02 lamp be connected to the service panel voltage?

_____ 3. For the given service, can a model number HID-03 lamp be connected to the service panel voltage?

_____ 4. For the given service, can a model number HID-04 lamp be connected to the service panel voltage?

_____ 5. For the given service, can a model number HID-05 lamp be connected to the service panel voltage?

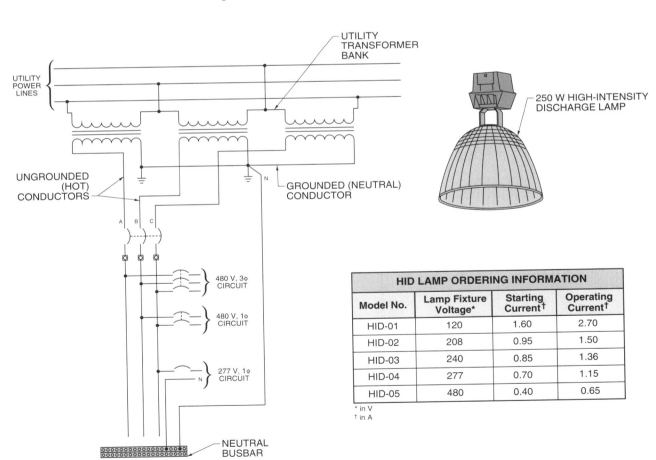

HID LAMP ORDERING INFORMATION			
Model No.	Lamp Fixture Voltage*	Starting Current†	Operating Current†
HID-01	120	1.60	2.70
HID-02	208	0.95	1.50
HID-03	240	0.85	1.36
HID-04	277	0.70	1.15
HID-05	480	0.40	0.65

* in V
† in A

277/480V, THREE-PHASE SYSTEMS

_____ 6. What is the advantage of using a lamp with a higher rated voltage when the lamp still only delivers 250 W, regardless of the voltage supply?

_____ 7. How many model number HID-04 lamps can safely be connected to a 15 A circuit breaker if the total operating current of the circuit does not exceed 85% of the circuit breaker rating?

_____ 8. How many model number HID-05 lamps can safely be connected to a 15 A circuit breaker if the total operating current of the circuit does not exceed 85% of the circuit breaker rating?

Activity 11-9: Plug and Receptacle Configurations and Ratings

A receptacle that has a current rating that is at least 20% higher than the maximum current draw of the load is used for each load. This ensures that the receptacle, branch circuit wire, and circuit breaker are at no greater than 80% capacity when the load is ON. For each of the electric heater loads listed below, determine the proper minimum NEMA receptacle configuration (5-15, 5-20, 7-50, etc.) using the Nonlocking Wiring Devices on page 158.

Load	Load Voltage Rating*	Load Current Rating†
1	120	28
2	240	28
3	120	8-13
4	277	17
5	120	3-6
6	240	8

* in V
† in A

_____ 1. NEMA receptacle configuration ___ is used for load 1.

_____ 2. NEMA receptacle configuration ___ is used for load 2.

_____ 3. NEMA receptacle configuration ___ is used for load 3.

_____ 4. NEMA receptacle configuration ___ is used for load 4.

_____ 5. NEMA receptacle configuration ___ is used for load 5.

Solid-State Devices and System Integration

12 Applications

Application 12-1: Electronic Device Symbols

Electronic Symbols

A s*ymbol* is a graphic element that represents a quantity or unit. Symbols can be used to conveniently represent solid-state components on electronic circuit diagrams. Symbols are used on electronic diagrams to identify the component used and how the components are interconnected. Each symbol must be understood in order to design, build, or troubleshoot an electronic circuit. **See Electronic Symbols in Appendix.**

Application 12-2: Digital Circuit Logic Functions

Digital Circuit Logic

A digital signal is a signal that is either high (ON) or low (OFF). A high signal in an electronic circuit is normally 5 VDC, but can be from 2.4 VDC to 5 VDC. A low signal is normally 0 VDC, but can be from 0 VDC to 0.8 VDC. Digital logic gates are used to control electronic circuits by making output decisions based on their inputs. The AND, OR, and NOT logic gates are the three basic logic functions that make up most digital circuit logic. The NOT gate is used to invert the output signal from a gate. The NOR gate is a NOT/OR (inverted OR) gate. The NAND gate is a NOT/AND (inverted AND) gate. AND, OR, NOT, NOR, and NAND logic have the same meaning for digital logic, hardwired electrical logic, and relay logic. **See Basic Logic Functions.** *Note:* The pushbuttons to all logic gates are normally connected to produce a low-voltage signal when the pushbuttons are open. The pushbuttons here are drawn to simplify their understanding.

BASIC LOGIC FUNCTIONS

Function	Digital Symbol	Description
AND	(AND gate diagram with 5 VDC, inputs, output)	**ENERGIZED** The output is energized if all inputs are activated. **DE-ENERGIZED** The output is de-energized if any one of the inputs is deactivated.
OR	(OR gate diagram with 5 VDC, inputs, output)	**ENERGIZED** The output is energized if one or more inputs are activated **DE-ENERGIZED** The output is de-energized if all of the inputs are deactivated.
NOT	(NOT gate diagram with 5 VDC, input, output)	**ENERGIZED** The output is energized if the input is not activated. **DE-ENERGIZED** The output is de-energized if the input is activated.
NOR	(NOR gate diagram with 5 VDC, inputs, output)	**ENERGIZED** The output is energized if none of the inputs are activated. **DE-ENERGIZED** The output is de-energized if one or more of the inputs is activated.
NAND	(NAND gate diagram with 5 VDC, inputs, output)	**ENERGIZED** The output is energized unless all inputs are activated. **DE-ENERGIZED** The output is de-energized if all the inputs are activated.

Digital AND Circuit Logic

An *AND gate* is a digital logic device that has two or more inputs and one output. The logic decision of an AND gate is based on the status of the inputs. If all inputs are high, the output is high. In all other cases, the output is low.

AND circuit logic is the same for hardwired electrical circuits using two or more normally open contacts connected in series and digital circuits using AND gates. A digital logic control circuit performs the same function as the electrical control circuit. The digital control circuit uses 5 VDC and the electrical control circuit uses 120 VAC for the control voltage. **See AND Logic.**

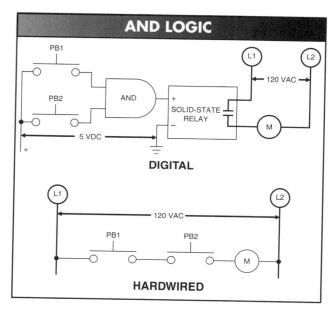

Digital OR Circuit Logic

An *OR gate* is a digital logic device that has two or more inputs and one output. The logic decision of an OR gate is based on the status of the inputs. If any of the inputs are high, the output is high. The output is low only when all inputs are low. OR circuit logic is the same for both digital circuits using OR gates and hardwired electrical circuits using two or more normally open contacts connected in parallel. **See OR Logic.**

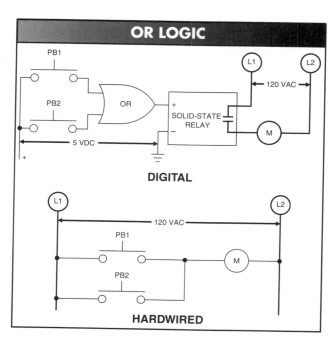

Digital NOT Circuit Logic

A *NOT gate* is a digital logic device that has one input and one output. The output of a NOT gate is the opposite of the input. If the input is low, the output is high. If the input is high, the output is low. NOT circuit logic is the same for both digital circuits using a NOT gate and hardwired electrical circuits using normally closed contacts. **See NOT Logic.** A circle placed at the output of a logic gate symbol indicates the output signal is inverted.

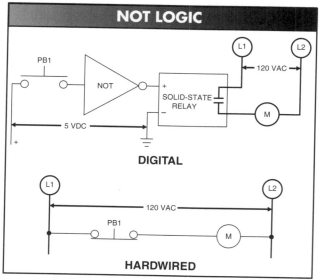

Digital NOR Circuit Logic

A *NOR gate* is a digital logic device that has two or more inputs and one output. The logic decision of a NOR gate is based on the status of the inputs. If one or all inputs are high (ON), the output is low (OFF). The output is high (ON) only when all inputs are low (OFF). NOR circuit logic is the same for both digital circuits using NOR gates and hardwired electrical circuits using two or more normally closed contacts connected in series. **See NOR Logic.**

Digital NAND Circuit Logic

A *NAND gate* is a digital logic device that has two or more inputs and one output. The logic decision of a NAND gate is based on the status of the inputs. If all inputs are high (ON), the output is low (OFF). In all other cases, the output is high (ON). NAND circuit logic is the same both for digital circuits using NAND gates and hardwired electrical circuits using two or more normally closed contacts connected in parallel. **See NAND Logic.**

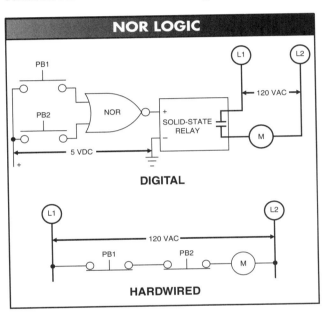
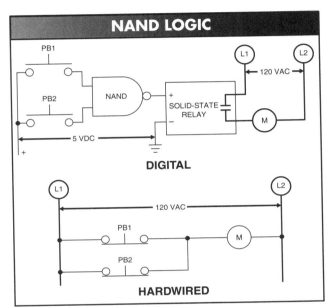

Application 12-3: Combination Logic Circuits

Combination Logic

Digital logic gates are combined to develop different logic functions. By combining the basic logic gates, digital circuits can solve almost any control application. This includes all basic hardwired industrial circuits such as start/stop, jogging, forward/reversing, etc. Digital circuits can also be used to solve switching applications that are not practical using hardwired circuits.

Start/Stop Control Circuits

A start/stop control circuit requires memory. In a memory circuit, the load (usually a motor starter) remains energized after a start pushbutton is pressed and released and remains energized until a stop pushbutton is pressed. The circuit has memory because the load remains energized after the start pushbutton is released. There must be a minimum flow of current for the circuit to stay ON. **See Start/Stop Control Circuit.**

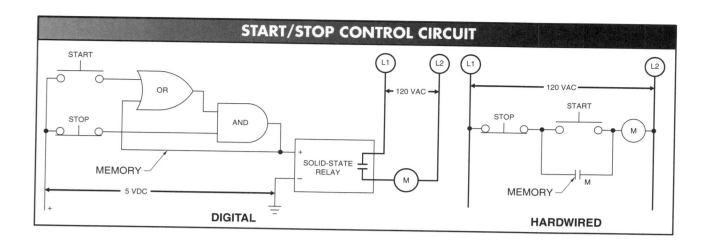

Jogging Control Circuits

Jogging is the frequent starting and stopping of a motor for short periods of time. In a jogging circuit, a motor is energized only when the jog pushbutton is held down. Jogging is used for positioning a load by moving the load a small distance each time the motor starts. **See Jogging Control Circuit.**

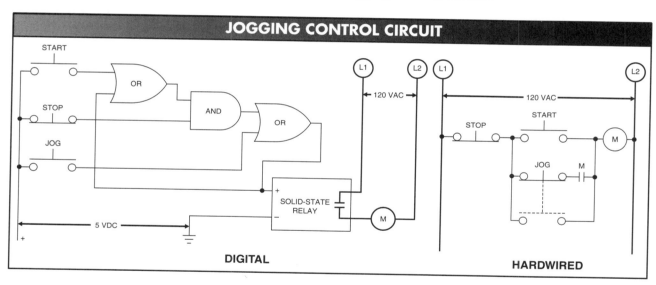

Coin Changer Circuits

AND and OR gates can be combined to produce the logic required in a basic coin changer. One set of inputs is activated as coins are inserted into a vending machine. Another set of inputs is activated when the cost of the product is determined. If the cost of the product selected is less than the coin inserted, the digital circuit makes the decision required to release the correct change. **See Coin Changer Circuit.** *Note:* To simplify this example, it is assumed that only one coin is inserted for the purchase of the product and the cost of the product selected is 5¢, 10¢, 15¢, 20¢, or 25¢.

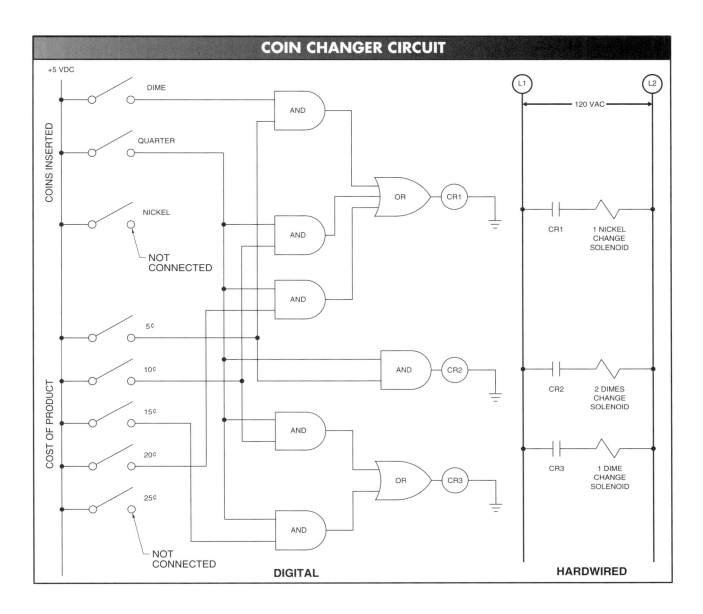

Application 12-4: Solid-State Relays and Switches

Digital System Interface

Because the output of a digital system is either 0 VDC or 5 VDC, an interface is required when operating a higher or different type of voltage. For example, a solid-state relay is used as an interface when digital circuits control DC outputs such as 12 VDC, 24 VDC, or 36 VDC solenoids, and AC outputs such as 120 VAC or 240 VAC motor starters. The voltage range allows a single solid-state relay to be used with most electronic circuits. However, the output of a solid-state relay is only used for AC or DC. **See Solid-State Relays.**

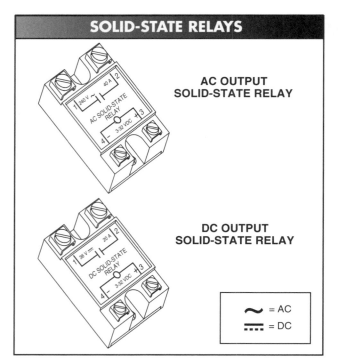

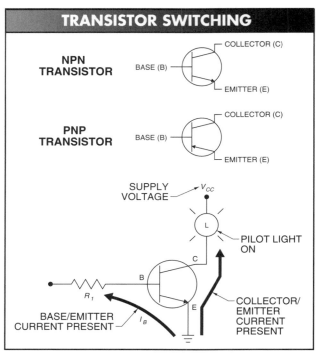

Transistor Switching

A *transistor* is a three-terminal device that controls current through the device depending on the amount of voltage applied to the base. Transistors can be used as either amplifiers or switches. Transistors may be either NPN or PNP transistors. An NPN transistor uses a small current and positive voltage at its base (relative to its emitter) to control a large current through its collector and emitter terminals. A PNP transistor uses a small base current and negative base voltage (relative to its emitter) to control a large emitter-to-collector current. **See Transistor Switching.**

SCR Switching

A *silicon controlled rectifier (SCR)* is a solid-state rectifier with the ability to rapidly switch high currents. The three electrodes of an SCR are the anode, cathode, and gate. The gate serves as the control point for the SCR. An SCR differs from a diode in that it does not pass significant current, even when forward biased, unless the anode voltage equals or exceeds the voltage point determined by the current applied to the gate. Each SCR is designed to turn ON the load-switching circuit at a predetermined voltage. SCRs switch direct current. **See SCR Switching.**

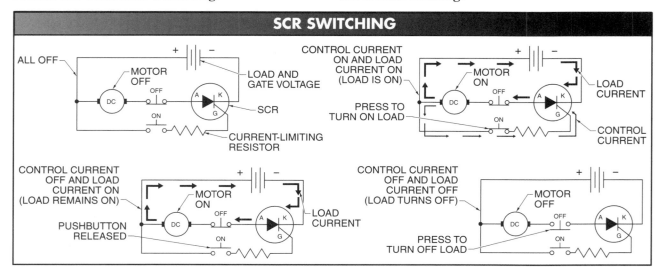

Triac Switching

A *triac* is a three-terminal semiconductor that is triggered into conduction in either direction by a small current at its gate. Triacs are suitable for AC applications because they are designed to pass current in both directions. **See Triac Switching.**

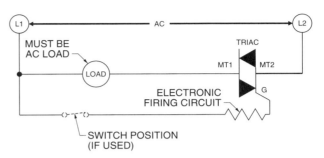

TRIAC SWITCHING

Application 12-5: Photovoltaic Cells

Solar Cells

A *photovoltaic cell (solar cell)* is a device that converts solar energy to electrical energy. Many different solar cells are available and are rated by the amount of energy they convert. Most manufacturers rate the output of a solar cell in volts and milliamps. To increase the current output, individual solar cells are connected in parallel.

The total current output is the sum of the current output of the individual cells, while the total voltage output is the same as the voltage output of an individual cell. To increase the voltage output, individual solar cells are connected in series. The total voltage output is the sum of the voltage output of the individual cells, while the total current output is the same as the current output of an individual cell.

Application 12-6: Troubleshooting Digital Circuits

Digital Logic Probes

When troubleshooting digital circuits, approximate test instrument readings should be anticipated if the test instrument readings are going to be used to help determine circuit problems. Digital logic circuits can be tested using a digital logic probe. A digital logic probe is a test instrument that is designed specifically for making basic tests on digital circuits. A digital logic probe usually only includes a few light emitting diode (LED) displays to show circuit operating condition. **See Digital Logic Probes.**

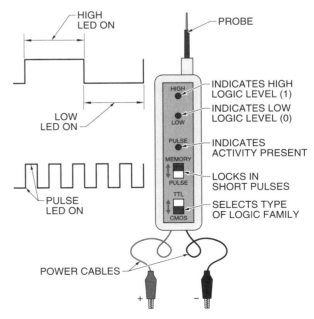

DIGITAL LOGIC PROBES

Typical displays on a digital logic probe include a logic high lamp, a logic low lamp, and a pulse lamp. The high lamp is ON when the tip of the logic probe detects a high (1) logic state. The low lamp is ON when the tip of the logic probe detects a low (0) logic state. The pulse lamp flashes when the tip of the logic probe detects logic activity present in the circuit. The pulse lamp also flashes when changes occur too fast to register with the high or low lamp.

When using a digital logic probe to troubleshoot a circuit, the power to the digital chips should be checked first. The circuit will not operate properly if the chip is not powered at all times. The positive input pin on the chip should indicate a logic high state and the negative pin should indicate a logic low state at all times. If the chip is powered, each individual gate on the chip should be checked for a logic high or logic low state. The condition of a pin (high or low) depends on the condition of the circuit (inputs ON or OFF) and the type of logic gate (AND, OR, etc.) being used.

Solid-State Devices and System Integration

12 Activities

Name _____ Date _____

Activity 12-1: Electronic Device Symbols

Draw the symbol for each electronic component.

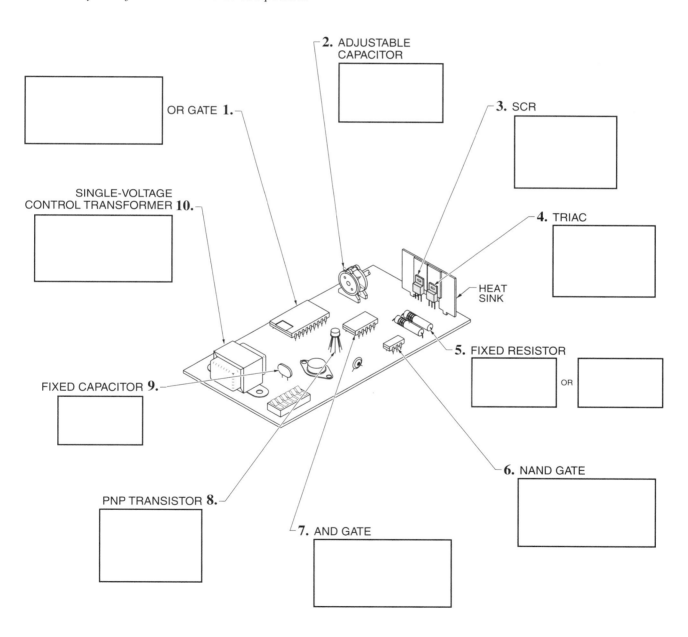

1. OR GATE
2. ADJUSTABLE CAPACITOR
3. SCR
4. TRIAC
5. FIXED RESISTOR   OR
6. NAND GATE
7. AND GATE
8. PNP TRANSISTOR
9. FIXED CAPACITOR
10. SINGLE-VOLTAGE CONTROL TRANSFORMER

HEAT SINK

Activity 12-2: Digital Circuit Logic Functions

In a car wash application, the electrical circuit includes several different types of circuit logic (AND, OR, NOT, NAND, NOR). The circuit logic can be developed by hard-wiring switches or by connecting the switches to digital logic gates that can be used to develop the circuit logic. Understanding both hardwired circuit logic and digital logic is required because both may be used in any given circuit.

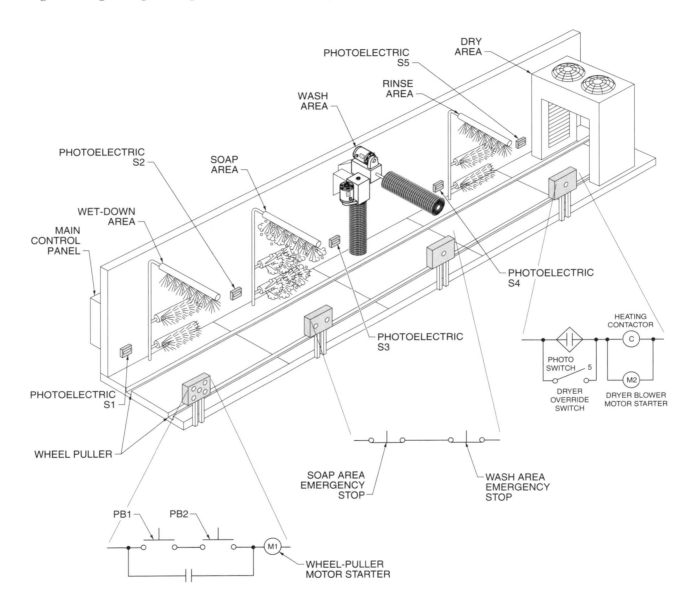

CAR WASH CIRCUIT

1. In the Car Wash Circuit, PB1 and PB2 are used to start the main wheel-puller motor. A digital logic gate could be used to produce the same circuit logic at this point in the circuit. Draw the digital logic gate that can be used to produce the same circuit logic and connect PB1 and PB2 to the gate inputs.

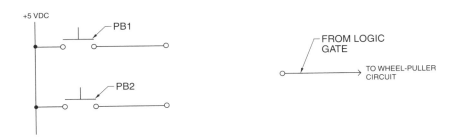

2. In the Car Wash Circuit, the soap area emergency stop and wash area emergency stop are used to manually stop the system if there is a problem at either location. A digital logic gate could be used to produce the same circuit logic at this point in the circuit. Draw the digital logic gate that can be used to produce the same circuit logic and connect each pushbutton to the gate inputs.

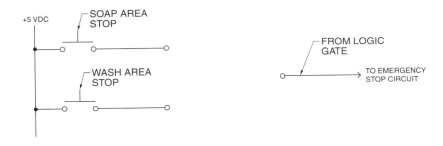

3. In the Car Wash Circuit, photo switch 5 and the dryer override switch are used to control the dry area. A digital logic gate could be used to produce the same circuit logic at this point in the circuit. Draw the digital logic gate that can be used to produce the same circuit logic and connect photo switch 5 and the dryer override switch to the gate inputs.

Activity 12-3: Combination Logic Circuits

In a digital circuit, a high logic level (5 VDC) can be represented by the number 1, and a low logic level (0 VDC) can be represented by the number 0. Using a 1 and 0 simplifies the circuit. For example, in Coin Changer Circuit 1, the top AND logic gate would have a zero placed at each input and output when a 25¢ coin is inserted in the machine and a 10¢ product is selected. If the 10¢ switch is open (dime not inserted) and the 5¢ switch is open (5¢ product not selected), both inputs would be low (0) because there is no voltage applied to the inputs. The output of an AND gate with no inputs present is also low (0).

1. For each of the seven remaining digital logic gates in Coin Changer Circuit 1, place a 1 (high circuit condition) or a 0 (low circuit condition) by the input and output of each gate based on the circuit condition. In addition, write the status (ON or OFF) above each solenoid based on the circuit condition.

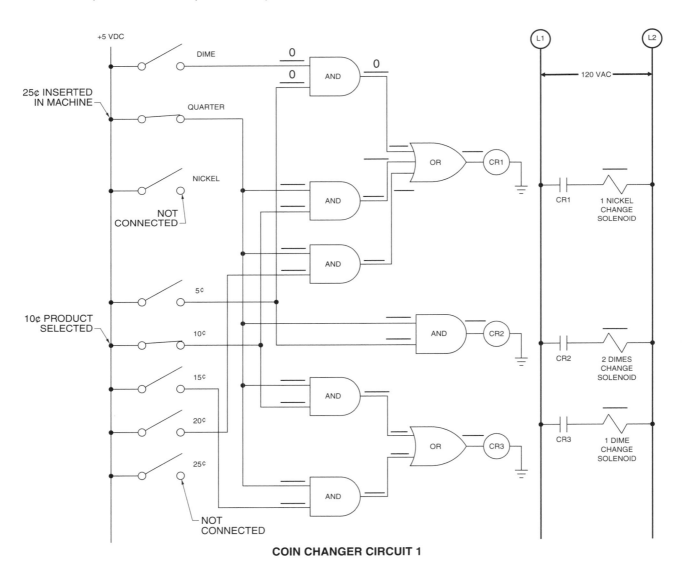

COIN CHANGER CIRCUIT 1

2. For each of the seven remaining digital logic gates in Coin Changer Circuit 2, place a 1 (high circuit condition) or a 0 (low circuit condition) by the input and output of each gate according to the circuit condition. In addition, write the status (ON or OFF) above each solenoid according to the circuit condition.

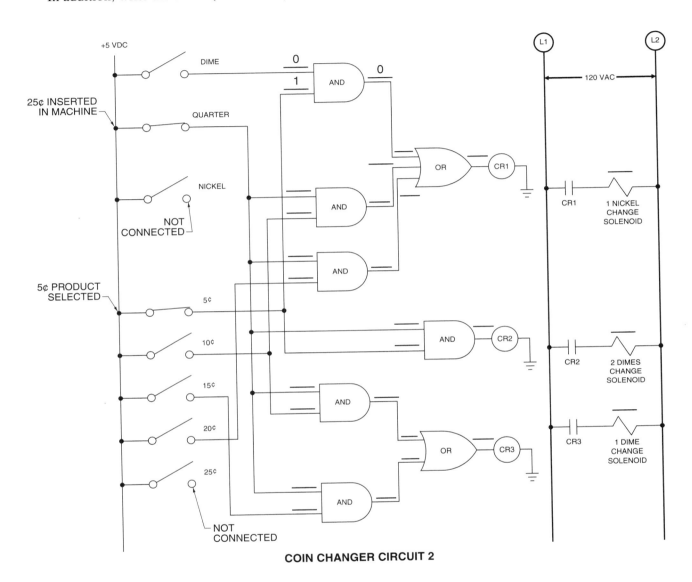

COIN CHANGER CIRCUIT 2

Activity 12-4: Solid-State Relays and Switches

1. Add a normally closed temperature switch into the circuit to control the PNP transistor turning ON and OFF. Connect the PNP transistor circuit to the solid-state relay so that the transistor controls the AC load.

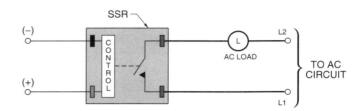

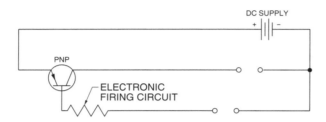

2. Add a normally open flow switch into the circuit to control the NPN transistor turning ON and OFF. Connect the NPN transistor circuit to the solid-state relay so that the transistor controls the AC load.

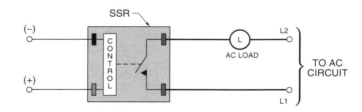

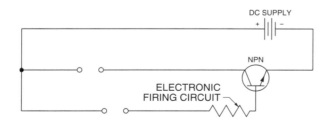

3. Add a normally open pushbutton into the circuit to control the SCR turning ON and OFF. Connect the SCR circuit to the DC compound motor so that the SCR controls the motor.

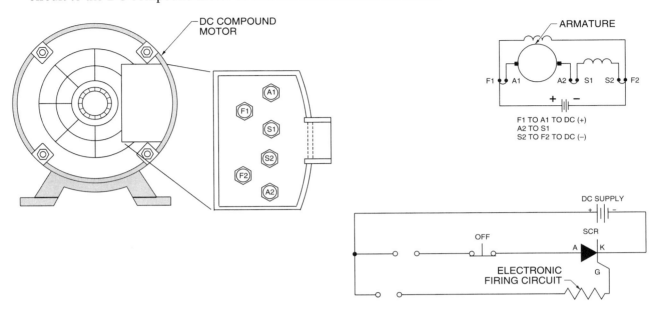

SCR SWITCHING

4. Add a normally open foot switch into the circuit to control the triac turning ON and OFF. Connect the triac circuit to the AC motor so that the triac controls the motor in the clockwise direction.

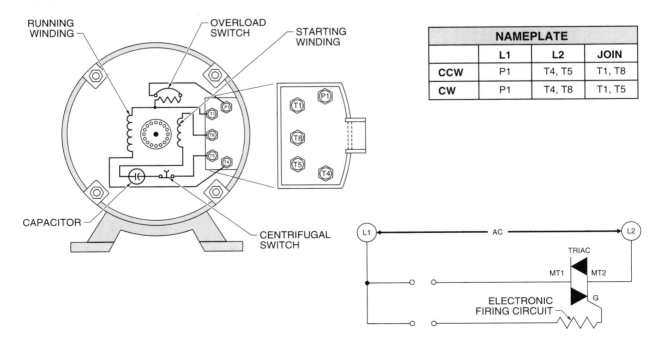

TRIAC SWITCHING

Activity 12-5: Photovoltaic Cells

Answer the following questions by referring to the weather satellite.

_____ 1. What is the total voltage output of circuit 1 if each solar cell is rated at 1 VDC?

_____ 2. What is the total current output of circuit 1 if each solar cell is rated at 40 mA?

_____ 3. What is the total power output of circuit 1 if each solar cell is rated 1 VDC at 40 mA?

_____ 4. What is the total voltage output of circuit 2 if each solar cell is rated at 1 VDC?

_____ 5. What is the total current output of circuit 2 if each solar cell is rated at 40 mA?

_____ 6. What is the total power output of circuit 2 if each solar cell is rated 1 VDC at 40 mA?

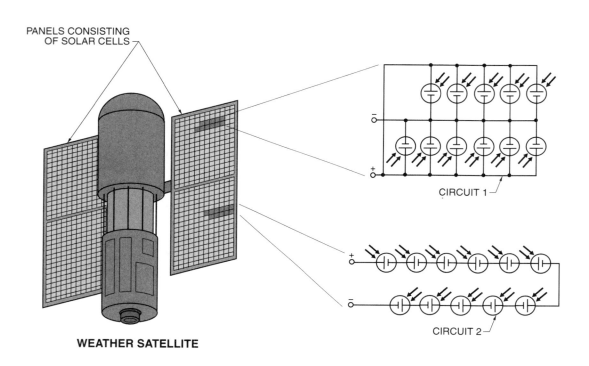

WEATHER SATELLITE

Activity 12-6: Troubleshooting Digital Circuits

Determine the expected logic probe reading (HIGH or LOW) for each logic probe if the circuit is operating properly.

_____ 1. The expected reading of logic probe 1 is ___.

_____ 2. The expected reading of logic probe 2 is ___.

_____ 3. The expected reading of logic probe 3 is ___.

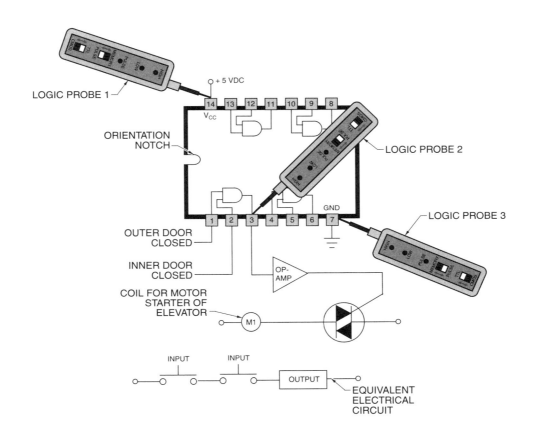

Timers and Counters

Applications 13

Application 13-1: ON-Delay Timer Applications

ON-Delay Timers

An *ON-delay timer* is a relay that provides a delay period after the relay is energized and before the relay contacts are switched. ON-delay timers are also referred to as operate delay relays. ON-delay timers are used in safety circuits that require a set time period after a switch is activated and before the load or loads are energized. The time delay provides safety by allowing time to review the decision or move out of the way of a process. ON-delay timers do not change position until the set time period passes after the timer receives power. Manufacturers provide specifications for ON-delay timers. **See Models A and B ON-Delay Timers.**

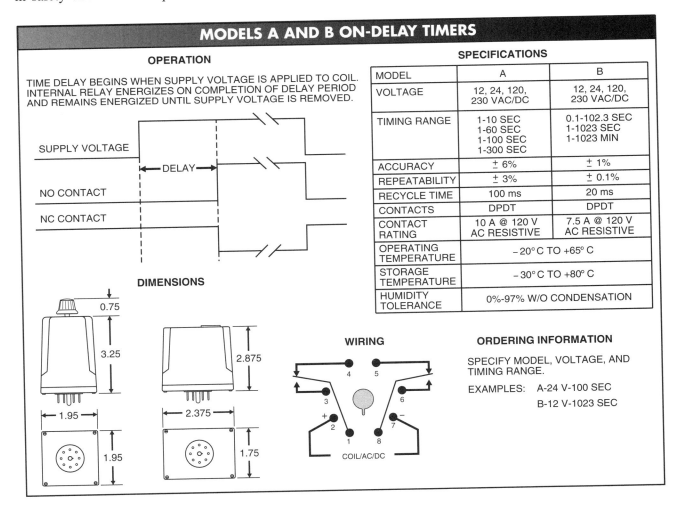

MODELS A AND B ON-DELAY TIMERS

OPERATION

TIME DELAY BEGINS WHEN SUPPLY VOLTAGE IS APPLIED TO COIL. INTERNAL RELAY ENERGIZES ON COMPLETION OF DELAY PERIOD AND REMAINS ENERGIZED UNTIL SUPPLY VOLTAGE IS REMOVED.

SPECIFICATIONS

MODEL	A	B
VOLTAGE	12, 24, 120, 230 VAC/DC	12, 24, 120, 230 VAC/DC
TIMING RANGE	1-10 SEC 1-60 SEC 1-100 SEC 1-300 SEC	0.1-102.3 SEC 1-1023 SEC 1-1023 MIN
ACCURACY	± 6%	± 1%
REPEATABILITY	± 3%	± 0.1%
RECYCLE TIME	100 ms	20 ms
CONTACTS	DPDT	DPDT
CONTACT RATING	10 A @ 120 V AC RESISTIVE	7.5 A @ 120 V AC RESISTIVE
OPERATING TEMPERATURE	–20° C TO +65° C	
STORAGE TEMPERATURE	–30° C TO +80° C	
HUMIDITY TOLERANCE	0%-97% W/O CONDENSATION	

ORDERING INFORMATION

SPECIFY MODEL, VOLTAGE, AND TIMING RANGE.

EXAMPLES: A-24 V-100 SEC
B-12 V-1023 SEC

Application 13-2: OFF-Delay Timer Applications

OFF-Delay Timers

An *OFF-delay timer* is a relay that provides a delay period after the relay is de-energized. OFF-delay timers are also referred to as release delay relays. OFF-delay timers are used in automatic garage door openers that require a time period in which a light is to remain ON after the door has opened or closed. Manufacturers provide specifications for OFF-delay timers. **See Models C and D OFF-Delay Timers.**

MODELS C AND D OFF-DELAY TIMERS

OPERATION

SUPPLY VOLTAGE IS CONSTANTLY APPLIED TO COIL. INTERNAL RELAYS ENERGIZE WHEN CONTROL SWITCH IS CLOSED. TIMING BEGINS WHEN CONTROL SWITCH IS OPENED. DELAY IS RESET BY RECLOSING CONTROL SWITCH. RELAY DE-ENERGIZES ON COMPLETION OF DELAY PERIOD.

SPECIFICATIONS

MODEL	C	D
VOLTAGE	24, 120 VAC/DC	12, 24, 120, 230 VAC/DC
TIMING RANGE	1-10 SEC 1-60 SEC 1-100 SEC 1-300 SEC	0.1-102.3 SEC 1-1023 SEC 1-1023 MIN
ACCURACY	± 6%	± 2%
REPEATABILITY	± 2.5%	± 0.1%
RECYCLE TIME	100 ms	20 ms
CONTACTS	DPDT	DPDT
CONTACT RATING	10 A @ 120 V AC RESISTIVE	7.5 A @ 120 V AC RESISTIVE
OPERATING TEMPERATURE	−25°C TO +65°C	
STORAGE TEMPERATURE	−30°C TO +90°C	
HUMIDITY TOLERANCE	0%-97% W/O CONDENSATION	

ORDERING INFORMATION

SPECIFY MODEL, VOLTAGE, AND TIMING RANGE.

EXAMPLES: C-24 V-10 SEC
                        D-120 V-1023 SEC

Application 13-3: One-Shot Timer Applications

One-Shot Timers

A *one-shot timer* is a relay that activates its contacts for a set time period after the relay is energized. One-shot timers are also referred to as pulse relays. One-shot timers are used in game machines to provide a time period in which the game can be played after the input is activated. Manufacturers provide specifications for one-shot timers. **See Models E and F One-Shot Timers.**

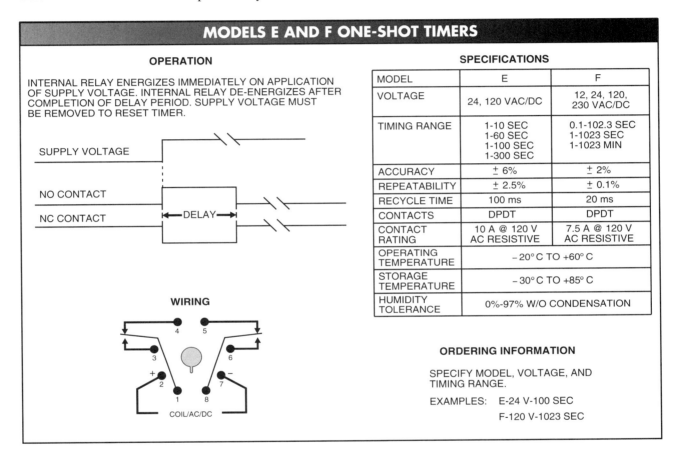

MODELS E AND F ONE-SHOT TIMERS

OPERATION

INTERNAL RELAY ENERGIZES IMMEDIATELY ON APPLICATION OF SUPPLY VOLTAGE. INTERNAL RELAY DE-ENERGIZES AFTER COMPLETION OF DELAY PERIOD. SUPPLY VOLTAGE MUST BE REMOVED TO RESET TIMER.

SPECIFICATIONS

MODEL	E	F
VOLTAGE	24, 120 VAC/DC	12, 24, 120, 230 VAC/DC
TIMING RANGE	1-10 SEC 1-60 SEC 1-100 SEC 1-300 SEC	0.1-102.3 SEC 1-1023 SEC 1-1023 MIN
ACCURACY	± 6%	± 2%
REPEATABILITY	± 2.5%	± 0.1%
RECYCLE TIME	100 ms	20 ms
CONTACTS	DPDT	DPDT
CONTACT RATING	10 A @ 120 V AC RESISTIVE	7.5 A @ 120 V AC RESISTIVE
OPERATING TEMPERATURE	−20° C TO +60° C	
STORAGE TEMPERATURE	−30° C TO +85° C	
HUMIDITY TOLERANCE	0%-97% W/O CONDENSATION	

ORDERING INFORMATION

SPECIFY MODEL, VOLTAGE, AND TIMING RANGE.

EXAMPLES: E-24 V-100 SEC
F-120 V-1023 SEC

Application 13-4: Recycle Timer Applications

Recycle Timers

A *recycle timer* is a relay that provides a repeating ON/OFF movement of the relay contacts after the timer is energized. Recycle timers are used in automatic lubrication control of motors and machines. Automatic lubrication is required to protect moving parts of motors and machines. For example, a recycle timer controls the time period between automatic lubrication for a band saw cutting operation. Manufacturers provide specifications for recycle timers. **See Models G and H Recycle Timers.**

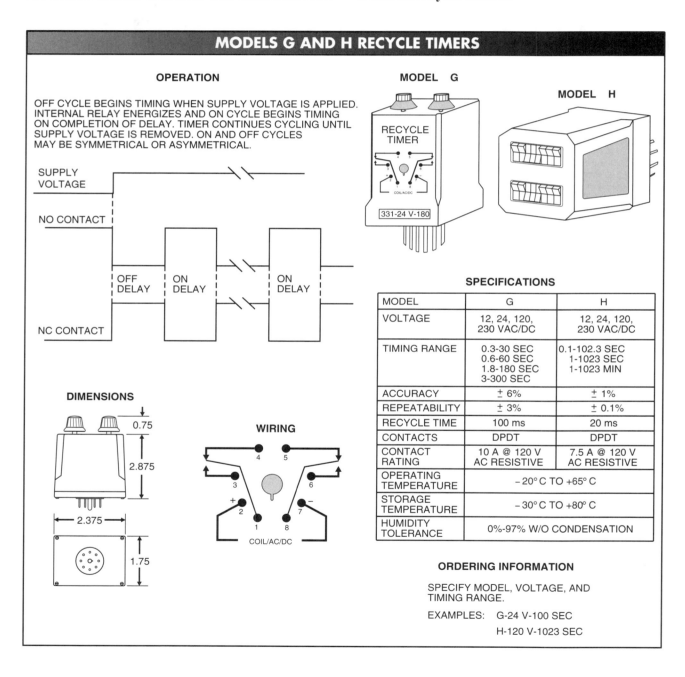

Application 13-5: Combination Timing Logic Applications

Combination Timing Logic

Industrial electrical applications often require multiple timing functions in the same circuit. For example, two different timers can be used to provide surge and backspin protection when starting and stopping a pump motor. In this application, an ON-delay timer is used to prevent starting surges from causing the pump to prematurely drop out during starting. A pump may drop out during starting because starting surges can cause the pressure switch to cycle open and closed. In addition, an OFF-delay timer can be used to prevent a backspin from immediately restarting the pump once the pump is stopped. This happens because the backspin caused by the pump turnoff could falsely activate the pressure switch. **See Surge and Backspin Protection.**

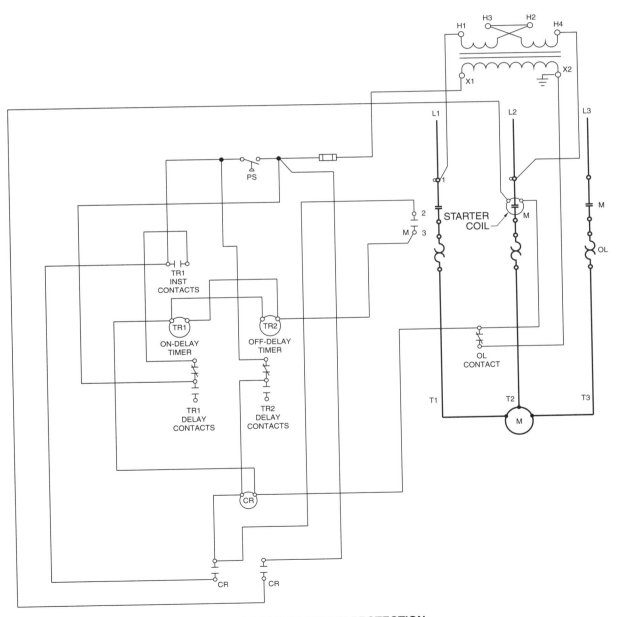

SURGE AND BACKSPIN PROTECTION

Application 13-6: Selecting and Setting Timers

Timer Selection

Multifunction timers are available for use in a variety of different applications. Multifunction timers use dual in-line package (DIP) switches to set the timing range and/or the timer type (ON-delay, one-shot, etc.). **See Multifunction Timer.**

For example, if DIP switches 1 and 2 are both placed in the up position and the control knob is set for 100% of the time range, the timer is set for 480 sec. If DIP switches 1 and 2 are both placed in the up position and the control knob is set for 50% of the time range, the timer is set for 240 sec (480 sec × 0.50 = 240 sec). Note that the minimum time setting on the 480 sec range is 24 sec. This means the timer cannot accurately be set below 24 sec on this range. However, when determining the timer setting using the control knob, only the maximum timer setting (15 sec, 60 sec, 480 sec, or 60 min) is used to determine the set time. Multifunction timers can also be programmed for special timing functions.

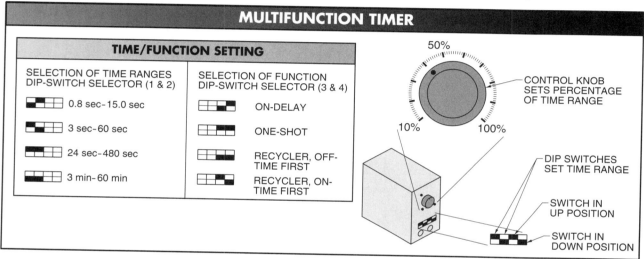

Application 13-7: Troubleshooting Timer Circuits

Timer Operating Condition

When troubleshooting a control circuit that includes a timer, the operating condition of the timer (timing or timed out) must be considered. This is because, even after the timer coil receives power, the timer contacts do not change state (normally open contacts close, and normally closed contacts open). Most timers include light-emitting diodes (LEDs) to help show the operating condition of the timer. **See Timer Operating Condition.**

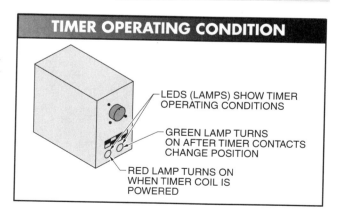

Timers usually include a single colored lamp that energizes any time the timer coil is powered. Timers may also include a second lamp of a different color that energizes when the timing contacts change state (timer timed out). On some single-lamp timers, the lamp may flash during timing to show that the timer is still timing and remain ON after the timer times out. Control circuit timers are normally solid-state devices.

Timers and Counters

13 Activities

Name _____ Date _____

Activity 13-1: ON-Delay Timer Applications

In a patient monitoring application, an ON-delay timer is used to monitor the patient's breathing rate. As long as the system ON/OFF switch is closed and the patient takes a breath before the timer times out, no warning is given. If the timer is allowed to time out (no patient breathing detected), the alarm automatically sounds. Answer the following questions using the Models A and B ON-Delay Timers chart on page 191 and the Patient Monitoring Application.

_____ 1. If a Model A or B timer is used in the application, which timer pin numbers are connected to wire reference numbers (terminal numbers) 3 and 2?

_____ 2. If a Model A or B timer is used in the application, which timer pin numbers are connected to wire reference numbers 1 and 6?

_____ 3. If a Model A timer is used in the application, what is the maximum current of a resistive load that can be switched?

_____ 4. If a Model A timer is used in the application, what is the minimum control circuit voltage with which a timer can be used?

_____ 5. If a Model B timer is used in the application, what is the maximum current of a resistive load that can be switched?

_____ 6. If a Model B timer is used in the application, what is the minimum control circuit voltage in which a timer can be used?

_____ 7. Which Model timer provides the shortest timing range that can be set?

_____ 8. Which Model timer provides the longest timing range that can be set?

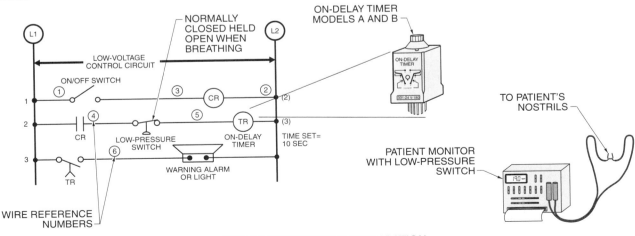

PATIENT MONITORING APPLICATION

197

Activity 13-2: OFF-Delay Timer Applications

In an emergency shower application, an OFF-delay timer is used to automatically sound an alarm for a set time every time anyone uses the emergency shower. Answer the following questions using the Models C and D OFF-Delay Timers chart on page 192 and the Emergency Shower Application.

_____ 1. If a Model C or D timer is used in the application, which timer pin numbers are connected to wire reference numbers (terminal numbers) 1 and 2?

_____ 2. If a Model C or D timer is used in the application, which timer pin numbers are connected to wire reference numbers 1 and 3?

_____ 3. If a Model C or D timer is used in the application, which timer pin numbers are connected to the start switch?

_____ 4. If a Model C timer is used in the application, what is the maximum current of a resistive load that can be switched?

_____ 5. If a Model C timer is used in the application, what is the minimum control circuit voltage with which a timer can be used?

_____ 6. If a Model D timer is used in the application, what is the maximum current of a resistive load that can be switched?

_____ 7. If a Model D timer is used in the application, what is the minimum control circuit voltage with which a timer can be used?

_____ 8. Which Model timer provides the shortest timing range that can be set?

_____ 9. Which Model timer provides the longest timing range that can be set?

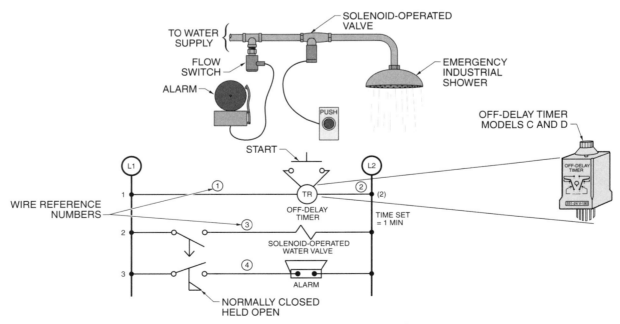

EMERGENCY SHOWER APPLICATION

Activity 13-3: One-Shot Timer Applications

In a plastic pallet wrap application, a one-shot timer is used to set the amount of wrap time in which the plastic is applied. Answer the following questions using the Models E and F One-Shot Timers chart on page 193 and the Pallet-Wrap Application.

_____ 1. If a Model E or F timer is used in the application, which timer pin numbers are connected to wire reference numbers (terminal numbers) 4 and 2?

_____ 2. If the control circuit is DC, which timer coil pin number is connected to the positive side of the control circuit?

_____ 3. If a Model E or F timer is used in the application, which timer pin numbers are connected to wire reference numbers 1 and 5?

_____ 4. Can either timer model be used in a 12 V control circuit?

_____ 5. Can either timer model be used in a 120 V control circuit?

_____ 6. Can either timer model be used in a 230 V control circuit?

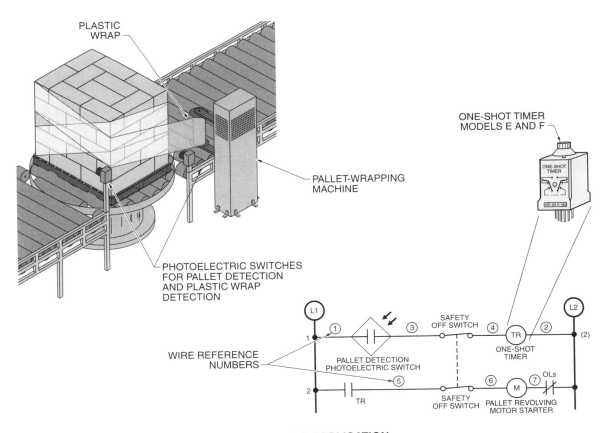

PALLET-WRAP APPLICATION

Activity 13-4: Recycle Timer Applications

In a product mixing application, a recycle timer is used to automatically keep the product mixed. The recycle timer determines the time the mixer is ON and the amount of time the mixer is OFF. Answer the following questions using the Models G and H Recycle Timers chart on page 194 and the Product Mixing Application.

_____ 1. If a Model G or H timer is used in the application, which timer pin numbers are connected to wire reference numbers 6 and 2?

_____ 2. If a Model G or H timer is used in the application, which timer pin numbers are connected to wire reference numbers 6 and 4?

_____ 3. If a Model G timer is used in the application, what is the maximum time for which the mixing motor could be OFF before automatically turning back ON when the selector switch is in the manual position?

_____ 4. If a Model H timer is used in the application, what is the maximum time for which the mixing motor could be OFF before automatically turning back ON when the selector switch is in the manual position?

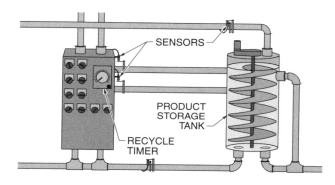

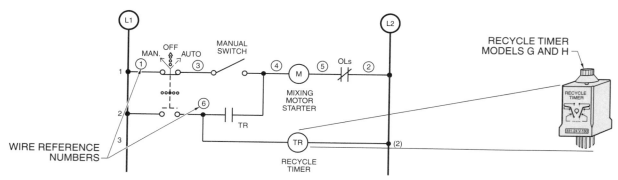

PRODUCT MIXING APPLICATION

Activity 13-5: Combination Timing Logic Applications

Having access to a control circuit line diagram is required for understanding how a circuit operates. The line diagram is used to help troubleshoot the circuit operation. Situations exist in which a circuit being tested has no line diagram available. When this occurs, following each of the wires in the wiring diagram can help in drawing the line diagram.

1. Complete the line diagram for the Surge and Backspin Protection wiring diagram on page 195. Use the correct ON-Delay and OFF-Delay timing symbols. The normally open (NO) instantaneous timer contact of TR1 is drawn the same as a standard normally open (NO) contact and marked TR1. Label every component in the circuit (CR, M, OLs, PS, etc.).

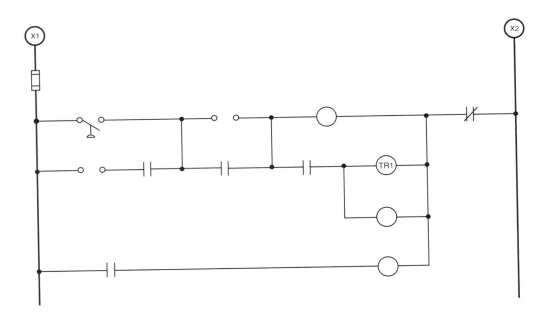

Activity 13-6: Selecting and Setting Timers

Determine which timer (timer 1, timer 2, or both) can be used in each application. If a timer cannot be used in the application, draw an X over the DIP switch for that timer. If a timer can be used, fill in the position of each DIP switch for that timer to meet the application requirements. Note: When selecting the time setting on a timer, set the timer for the lowest time range available that still falls within the timing application range.

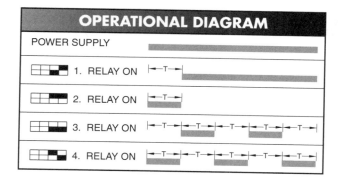

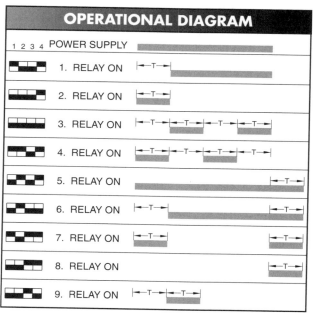

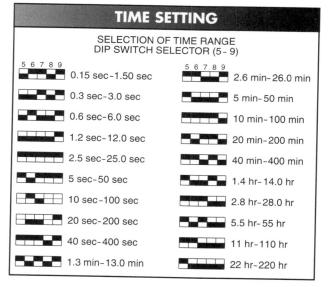

1. An ON-delay timer is used to sound an alarm if there is no flow for 10 sec or more.

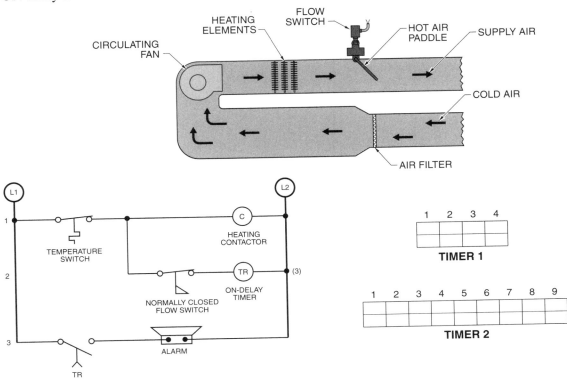

2. An OFF-delay timer is used to keep a solenoid ON when the start pushbutton is pressed and for 20 sec after the start pushbutton is released.

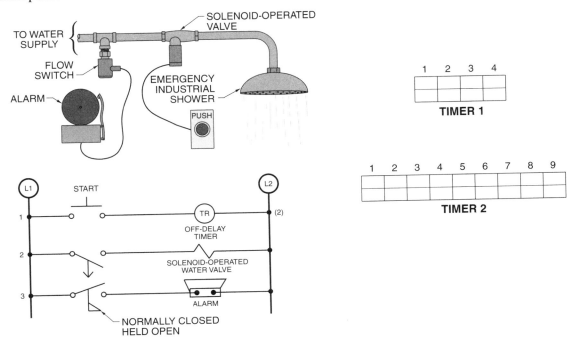

3. A one-shot timer is used to turn ON a pallet rotating motor starter when a photoelectric switch detects a pallet. The timer keeps the starter ON for 35 sec, even when the photoelectric switch still detects the pallet.

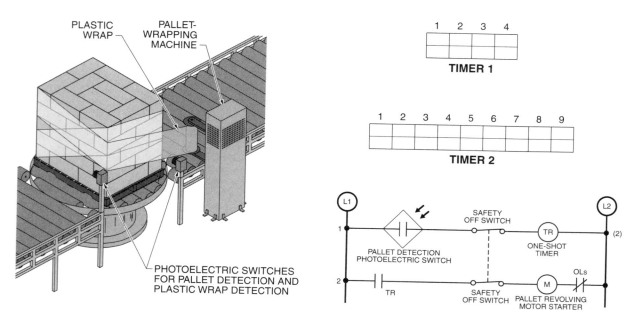

4. A recycle timer (with OFF-delay first) is used to mix product every 5 min, 30 sec.

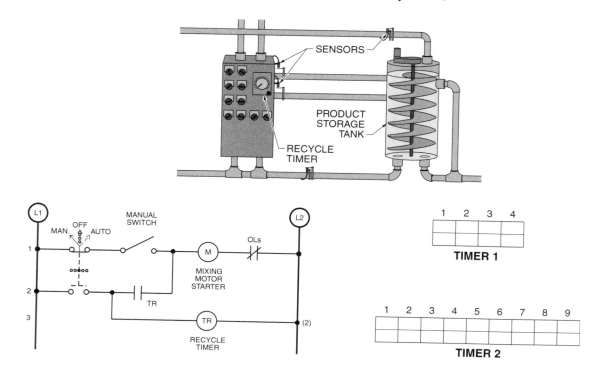

Activity 13-7: Troubleshooting Timer Circuits

When troubleshooting recycle timer circuits, approximate meter readings should be anticipated if the meter readings are going to be used to help determine circuit problems. Determine the expected DMM readings if the circuit is operating properly.

_____ 1. The expected reading of DMM 1 with the selector switch in the auto position and the motor OFF is ___ VAC.

_____ 2. The expected reading of DMM 1 with the selector switch in the auto position and the motor ON is ___ VAC.

_____ 3. The expected reading of DMM 2 with the selector switch in the auto position and the motor OFF is ___ VAC.

_____ 4. The expected reading of DMM 2 with the selector switch in the auto position and the motor ON is ___ VAC.

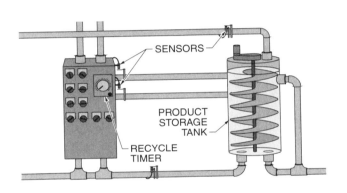

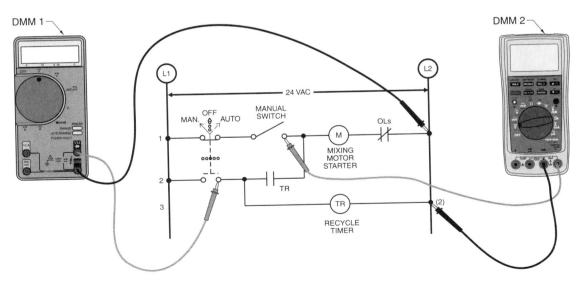

When troubleshooting one-shot timer circuits, approximate meter readings should be anticipated if the meter readings are going to be used to help determine circuit problems. Determine the expected DMM readings if the circuit is operating properly.

_____ 5. The expected reading of DMM 1 when a pallet is first detected in place on the assembly line is ___ VAC.

_____ 6. The expected reading of DMM 1 when no pallet is present on the assembly line is ___ VAC.

_____ 7. The expected reading of DMM 2 when the wrap cycle is complete on the assembly line is ___ VAC.

_____ 8. The expected reading of DMM 2 when no pallet is present on the assembly line is ___ VAC.

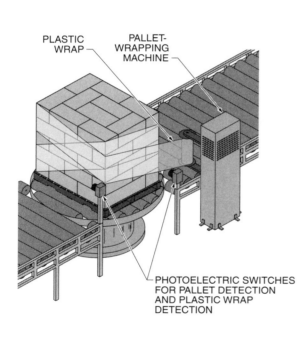
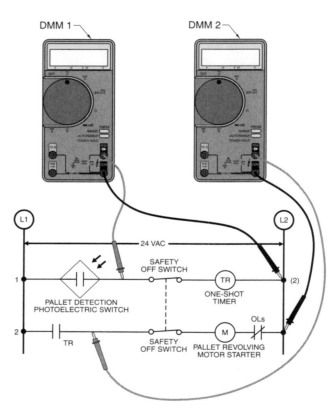

Relays and Solid-State Starters

Applications 14

Application 14-1: Relays

Relay Contact Identification

The different contact arrangements of relays and timers are shown in schematic diagrams that describe the poles, throws, and breaks of the relay or timer. Electricians must identify contacts when installing, replacing, or specifying relays. A *contact* is the conducting part of a switch that operates with another conducting part of the switch to make or break a circuit. A *break* is the number of separate places on a contact that open or close an electrical circuit. Contacts are either single-break or double-break. A *pole* is the number of completely isolated circuits that a relay can switch. A single-pole contact can carry current through only one circuit at a time. A double-pole contact can carry current through two circuits simultaneously. A *throw* is the number of closed contact positions per pole. A single-throw contact can control only one circuit. A double-throw contact can control two circuits. **See Relay Contacts.**

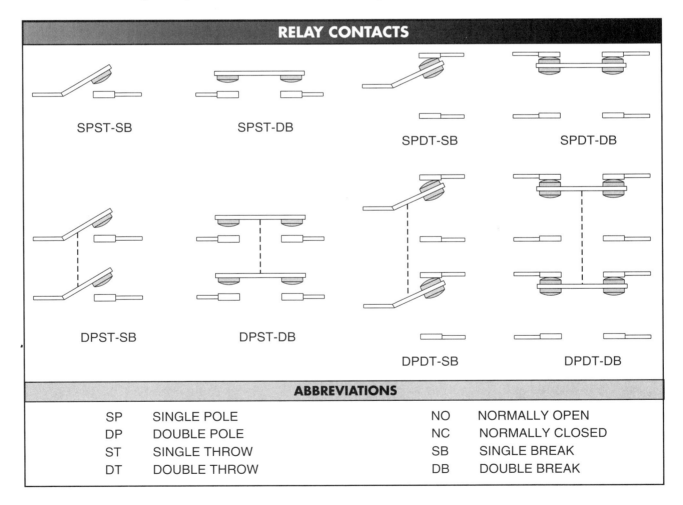

Relay Applications

Relays, like all electrical switches, are used in applications that require an electrical circuit to be opened and closed. The main difference between an electrical circuit containing a switch and an electrical circuit containing a relay is that when a relay is used, the control switch (pressure switch, temperature switch, etc.) controls the energizing or de-energizing of the relay coil. The relay contacts switch the circuit load (lamps, heating elements, solenoids, etc.). The advantage of relay contacts switching a load is that the control switch only has to carry the low voltage (12 V, 24 V, etc.) and low current (in mA) of the relay coil. The relay contacts carry the high voltage (115 V, 230 V, etc.) and high load current (in A).

General-Purpose Relays

Manufacturers use numbers to identify the different types of relays. These numbers include the dimensions, electrical and mechanical features, and contact arrangements of the relay. When ordering relays from manufacturers, understanding relay terminology is required. For example, a model 01-B-1-B-1A-120A is similar to a model number 01-B-1-B-2A-120A. However, the second model number is a relay with twice the number of available contacts (DPST) as the first model (SPST). **See Relay Specifications.**

Latching Relays

Control circuit memory is developed by using holding contacts connected in parallel with the control circuit start pushbutton, or by using a latching relay. When using holding contacts connected in parallel with the start pushbutton, the load is energized when the start pushbutton is pressed and released. The load remains energized until the memory circuit is broken. The memory circuit is normally broken by a stop pushbutton. The memory circuit is also broken when a voltage failure occurs. In this case, the circuit does not turn ON when the voltage returns. This circuit is referred to as a low-voltage (undervoltage) protection circuit because the memory circuit is broken when a low-voltage condition occurs. When power returns, the start pushbutton must be pressed to energize the load and memory circuit.

A *latching relay* is a relay that maintains a given contact position by means of a mechanical latch. When using a latching relay to develop circuit memory, the latching relay provides a holding function through a set/reset coil on the relay. The latching relay contacts change position when voltage is applied to the set coil. The contacts stay in this position even if the voltage is removed from the relay. To change the contacts position, voltage must be applied to the latching relay reset coil. **See Latching Relay Wiring Diagrams.**

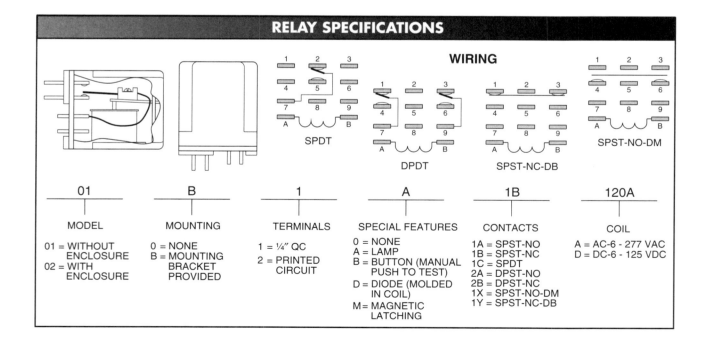

LATCHING RELAY WIRING DIAGRAMS

Application 14-2: Heat Sink Selection

Heat Sinks

Solid-state relays use heat sinks to prevent damage to the switching element of the relay. The size of a heat sink is determined by dividing the temperature rise at the relay by the power at the relay contact. This ratio is the required thermal resistance of the heat sink (in °C/W) and is used when rating and selecting heat sinks. To select a heat sink for a relay, apply the following procedure:

1. Determine the current the relay controls. If the current is not known, apply Ohm's law.
2. Determine the ambient temperature of the relay installation.
3. Determine the temperature rise at the relay. To calculate temperature rise, apply the following formula:

$$T_R = M_{TR} - A_T$$

where
T_R = temperature rise at relay (in °C)
M_{TR} = maximum temperature rise at relay (normally 110°C)
A_T = ambient temperature at relay (in °C)

4. Determine the power drop at relay contacts. To calculate power drop, apply the following formula:

$$P_D = I_L \times V_D$$

where
P_D = power drop at relay contacts (in W)
I_L = load current (in A)
V_D = voltage drop at relay contacts (normally 1.6 V)

5. Determine the thermal resistance of the heat sink. The thermal resistance of a heat sink is calculated or found on relay specifications. To calculate thermal resistance, apply the following formula:

$$R_{TH} = \frac{T_R}{P_D}$$

where
R_{TH} = thermal resistance of heat sink (in °C/W)
T_R = temperature rise at relay (in °C)
P_D = power drop at relay contacts (in W)

6. Select the heat sink with a value equal to or smaller than the required size. **See Heat Sink Selections.**

HEAT SINK SELECTIONS

Type	H × W × L (mm)	R_{TH}*
01	15 × 79 × 100	2.5
02	15 × 100 × 100	2.0
03	25 × 97 × 100	1.5
04	37 × 120 × 100	0.9
05	40 × 160 × 150	0.5
06	40 × 200 × 150	0.4

*in °C/W

Example: Heat Sink Selection—Calculation

A relay controls a 25 A load in a 50°C ambient temperature installation. Find the required heat sink.

1. Determine the current the relay controls.

   The relay controls a 25 A load.

2. Determine the ambient temperature of the relay installation.

   The relay is installed in a 50°C ambient temperature location.

3. Determine the temperature rise at the relay.

   $T_R = M_{TR} - A_T$
   $T_R = 110 - 50$
   $T_R = 60°C$

4. Determine the power drop at relay contacts.

   $P_D = I_L \times V_D$
   $P_D = 25 \times 1.6$
   $P_D = 40\ W$

5. Determine the thermal resistance of the heat sink.

   $R_{TH} = \dfrac{T_R}{P_D}$
   $R_{TH} = \dfrac{60}{40}$
   $R_{TH} = 1.5°C/W$

6. Select a heat sink with a thermal resistance value equal to or smaller than the required size (from Heat Sink Selections). Minimum selection for 1.5°C/W is **Type 03**. Best selection is **Type 04**.

Heat Sink Selection—Relay Specifications

The thermal resistance of a heat sink can also be found using manufacturer relay specifications. To find the thermal resistance of a heat sink using manufacturer relay specifications, find the intersection of the load current and ambient temperature on a relay specification. The thermal resistance value is represented by the °C/W line directly above the intersection of the load current and ambient temperature. **See 40 A Relay Specifications.**

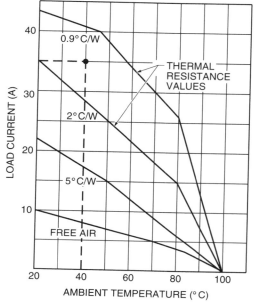

40 A RELAY SPECIFICATIONS

Example: Heat Sink Selection—Relay Specifications

A 40 A relay controls a 35 A load in a 40°C ambient temperature installation. Find the required heat sink.

The thermal resistance of the heat sink is found at the intersection of the load current and ambient temperature on a relay specification graph. The intersection of the 35 A and 40°C lines on the relay specification graph falls below the 0.9°C/W thermal resistance value. The minimum heat sink type with a value equal to or smaller than the required size (from Heat Sink Selections) is **Type 04**. The best selection is **Type 05**.

Application 14-3: Heat Sink Installation

Mounting Heat Sinks

A heat sink must be properly mounted and placed in a system to ensure proper heat transfer from the heat sink to the surrounding air. Proper heat sink mounting includes mounting the fins of the heat sink vertically to ensure the heated air is moved away from the relay. Proper placement includes mounting the heat sink to permit free airflow from natural convection or forced airflow. **See Mounting Heat Sinks.**

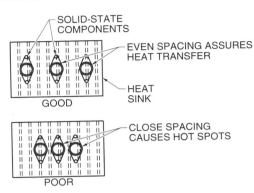

MOUNTING HEAT SINKS

Application 14-4: Solid-State Relay Installation

Solid-State Relays

Solid-state relays should be installed to allow maximum heat transfer from the relay. Maximum heat transfer depends on how well the heat produced inside a relay is transferred to the heat sink, how well the heat transferred to the heat sink is transferred to the surrounding air, and how well the surrounding air is replaced with cooler air.

Thermal conductive compound is a thermal paste used to reduce the high thermal resistance of the air gap between a heat sink and a solid-state relay. Thermal conductive compound is spread evenly over the entire surfaces to be joined. **See Thermal Conductive Compound.**

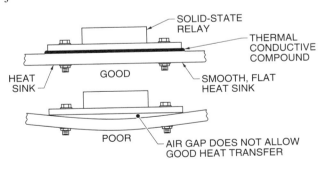

THERMAL CONDUCTIVE COMPOUND

Minimum Holding Current

The current draw of a load must be sufficient to keep a solid-state relay operating when the relay is conducting (load ON). *Minimum holding current* is the minimum current that ensures proper operation of a solid-state relay. Minimum holding current values range from 2 mA to 20 mA.

Operating current and minimum holding current values are not a problem when a solid-state relay controls low-impedance loads such as motor starters, relays, and solenoids. Operating current and minimum holding current values are a problem when a relay controls high-impedance loads such as programmable controllers and other solid-state devices. The operating current may be high enough to affect the load when the relay is not conducting. A load resistor must be connected in parallel with the load to correct this problem. The load resistor acts as an additional load that increases the total current in the circuit. Load resistors range in value from 4.5 kΩ to 7 kΩ. A 5 kΩ, 5 W resistor is used in most applications. **See Load Resistor.**

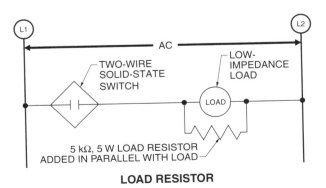

LOAD RESISTOR

Application 14-5: Troubleshooting at Motor Starters

Motor Starters

A motor starter is used to control the turning ON and OFF of the power (high-voltage) circuit of a motor. A motor starter may be a magnetic motor starter, electric motor drive, or solid-state starter. When troubleshooting at a motor starter, expected results should be anticipated. On the power supply side of the starter, the supply voltage should be present across any two of the power lines in a 3φ system. The actual supply voltage may be +5% to −10% of the rating of the load (motor, heating elements, etc.). However, once the supply voltage is determined (458 V, 459 V, 460 V, 461 V, etc.), the voltage between any two power lines should be within 3% of each other. When the load is turned ON, the voltage drop must be less than 3%. A portion of the voltage drop is caused by voltage being dropped across the starter, the voltage drop from conductors leading to the motor, and the voltage drop due to wire splices and terminations. The exact voltage drop across the starter can be measured by connecting a DMM set to measure voltage across individual mechanical or solid-state contacts (L1 to T1, L2 to T2, and L3 to T3). The voltage drop across individual contacts should be less than 3% of the supply voltage when the motor is running. There is no voltage drop across individual contacts when the motor is OFF. **See Troubleshooting at Motor Starter.**

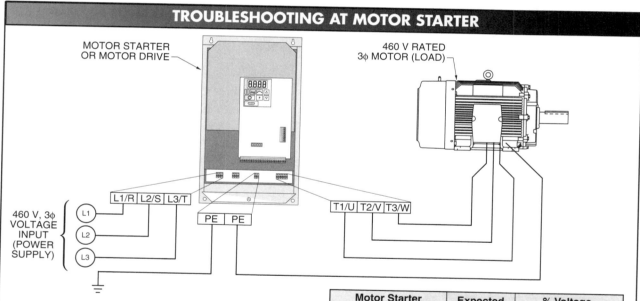

Power Supply DMM Measurement	Expected Results	% Voltage Drop
L1 – L2 (Motor OFF)	460 V	
L2 – L3 (Motor OFF)	460 V	+ 5% to − 10% of load's rating
L3 – L1 (Motor OFF)	460 V	
L1 – L2 (Motor ON)	460 V	
L2 – L3 (Motor ON)	460 V	3% or less of voltage when motor is OFF
L3 – L1 (Motor ON)	460 V	

Motor Starter DMM Measurement	Expected Results	% Voltage Drop
T1 – T2 (Motor OFF)	0 V	
T2 – T3 (Motor OFF)	0 V	0%
T3 – T1 (Motor OFF)	0 V	
T1 – T2 (Motor ON)	460 V	
T2 – T3 (Motor ON)	460 V	+ 5% to − 10% of load's rating
T3 – T1 (Motor ON)	460 V	
L1 – L2 (Motor OFF)	0 V	
L1 – L3 (Motor OFF)	0 V	0%
L1 – T1 (Motor OFF)	0 V	
L1 – T2 (Motor ON)	460 V	3% or less of voltage when motor is OFF
L1 – T3 (Motor ON)	460 V	
L1 – T1 (Motor ON)	0 V	0%

Relays and Solid-State Starters

14 Activities

Name ______________________ Date ______________

Activity 14-1: Relays

Use Arrangement 1 to answer questions 1 – 6.

______________ **1.** The number of poles is ___.

______________ **2.** The number of throws is ___.

______________ **3.** The number of breaks is ___.

______________ **4.** The coil pins are ___ and ___.

______________ **5.** The number of normally open contacts is ___.

______________ **6.** The number of normally closed contacts is ___.

2
5
7
A
(–)
B
(+)
NOTE: THE (–) AND (+) SIGNS ARE ADDED TO TERMINALS A AND B FOR DC RELAY MODELS

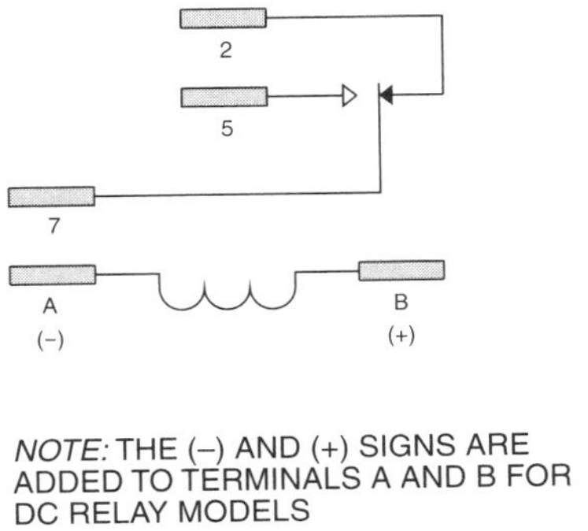

ARRANGEMENT 1

Use Arrangement 2 to answer questions 7 – 12.

______________ **7.** The number of poles is ___.

______________ **8.** The number of throws is ___.

______________ **9.** The number of breaks is ___.

______________ **10.** The coil pins are ___ and ___.

______________ **11.** The number of normally open contacts is ___.

______________ **12.** The number of normally closed contacts is ___.

1
3
4
6
7
9
A
(–)
B
(+)

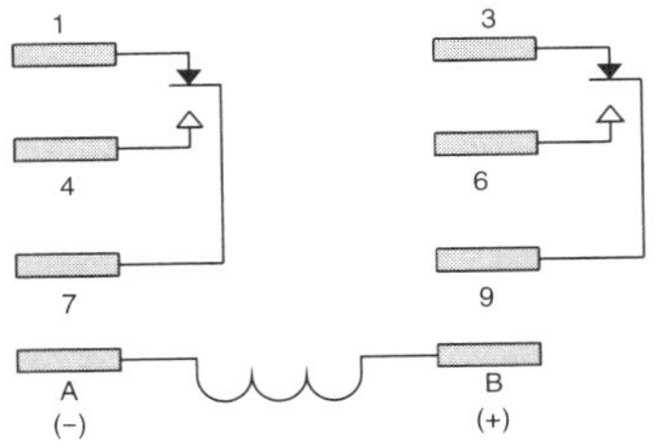

ARRANGEMENT 2

Use Arrangement 3 to answer questions 13 – 18.

______________ **13.** The number of poles is ___.

______________ **14.** The number of throws is ___.

______________ **15.** The number of breaks is ___.

______________ **16.** The coil pins are ___ and ___.

______________ **17.** The number of normally open contacts is ___.

______________ **18.** The number of normally closed contacts is ___.

1
2
3
4
5
6
7
8
9
A
(–)
B
(+)

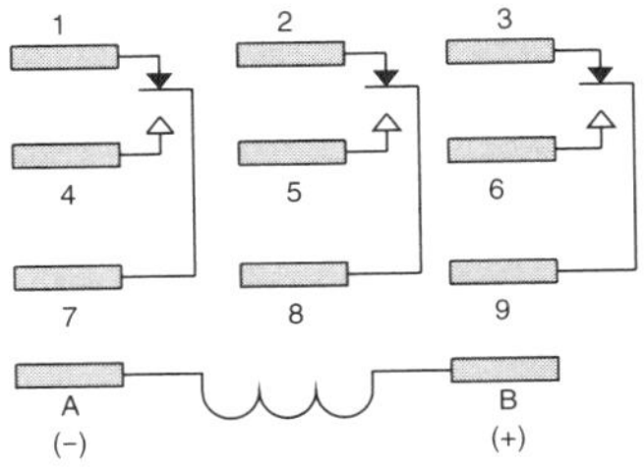

ARRANGEMENT 3

Use Arrangement 4 to answer questions 19–24.

_____ 19. The number of poles is ___.

_____ 20. The number of throws is ___.

_____ 21. The number of breaks is ___.

_____ 22. The coil pins are ___ and ___.

_____ 23. The number of normally open contacts is ___.

_____ 24. The number of normally closed contacts is ___.

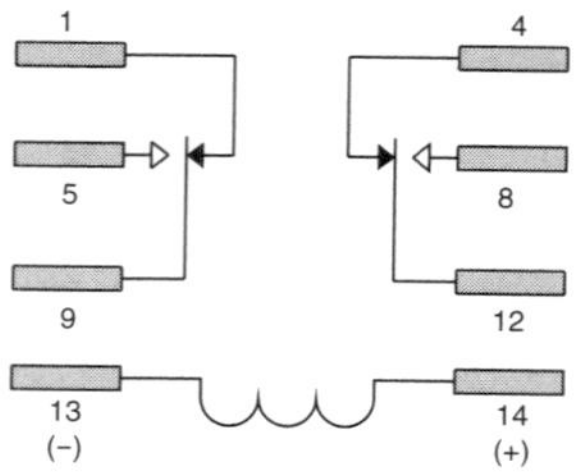

ARRANGEMENT 4

Use Arrangement 5 to answer questions 25–30.

_____ 25. The number of poles is ___.

_____ 26. The number of throws is ___.

_____ 27. The number of breaks is ___.

_____ 28. The coil pins are ___ and ___.

_____ 29. The number of normally open contacts is ___.

_____ 30. The number of normally closed contacts is ___.

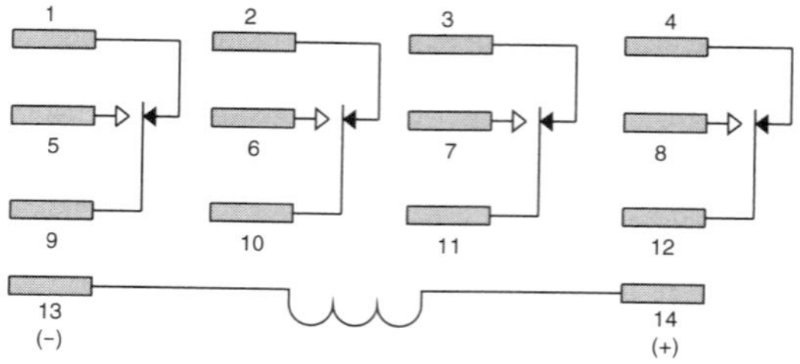

ARRANGEMENT 5

Use Arrangement 6 to answer questions 31–36.

_____ 31. The number of poles is ___.

_____ 32. The number of throws is ___.

_____ 33. The number of breaks is ___.

_____ 34. The coil pins are ___ and ___.

_____ 35. The number of normally open contacts is ___.

_____ 36. The number of normally closed contacts is ___.

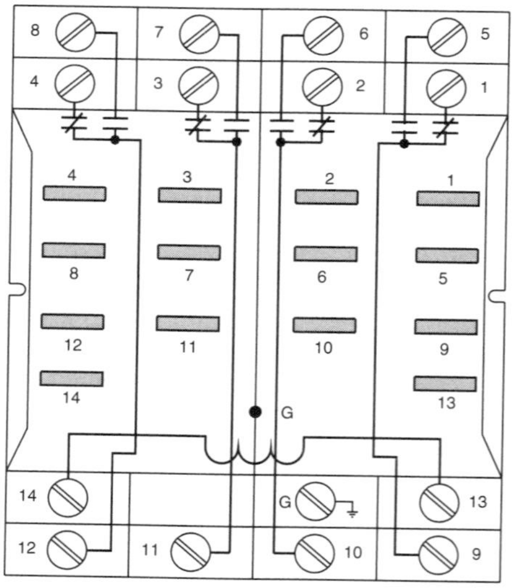

ARRANGEMENT 6

When selecting a relay type for an application, the relay contact arrangement must comply with the circuit requirement. In the following circuit, timers and relays (both double- and single-break) are required. The normally closed (NC) contact used in line 4 with TR1 is an example of a single-break contact. The instantaneous (INST) contact used in line 3 with TR1 is an example of a double-break contact. For each of the three relays, list the minimum number of contacts required for the application. Use Relay Contacts on page 207 to identify the arrangement type.

_____ **37.** The relay contact arrangement for coil C1 is ___.

_____ **38.** The relay contact arrangement for coil C2 is ___.

_____ **39.** The relay contact arrangement for coil C3 is ___.

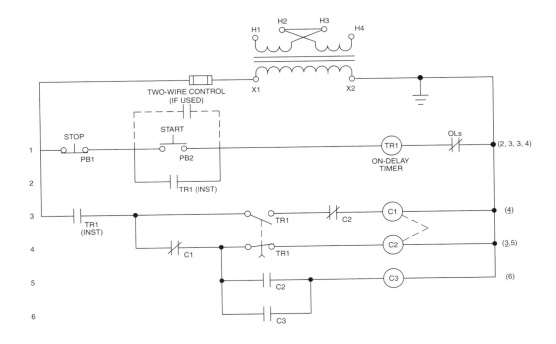

Answer the questions using the Relay Specifications chart on page 208. A relay has a model number 01-B-1-B-1C-120A.

_____ **40.** The relay has ___ normally open contacts.

_____ **41.** The terminal numbers of the contacts are ___.

_____ **42.** The relay has ___ normally closed contacts.

_____ **43.** The terminal numbers of the contacts are ___.

_____ **44.** The terminal numbers of the coil are ___.

_____ **45.** The rating of the coil is ___ VAC.

Activity 14-2: Heat Sink Selection

Select the correct size heat sink for each application using the Heat Sink Selections chart on page 210.

_____ 1. A 120 V, 30 A load is controlled by a relay installed in an ambient temperature of 40°C. A type ___ heat sink is required.

_____ 2. A 120 V, 8 Ω load is controlled by a relay installed in an ambient temperature of 75°C. A type ___ heat sink is required.

_____ 3. A 240 V, 2500 W load is controlled by a relay installed in an ambient temperature of 55°C. A type ___ heat sink is required.

_____ 4. A 240 V, 10 Ω load is controlled by a relay installed in an ambient temperature of 45°C. A type ___ heat sink is required.

_____ 5. A 240 V, 9.6 Ω, 6000 W load is controlled by a relay installed in an ambient temperature of 60°C. A type ___ heat sink is required.

Activity 14-3: Heat Sink Installation

Mark each relay/heat sink mounting method as correct or incorrect.

_____ 1. Heat sink mounting method is ___.

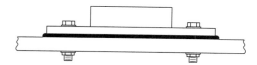

_____ 2. Heat sink mounting method is ___.

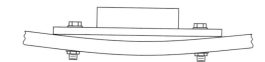

_____ 3. Heat sink mounting method is ___.

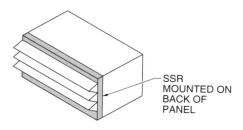

SSR MOUNTED ON BACK OF PANEL

_____ 4. Heat sink mounting method is ___.

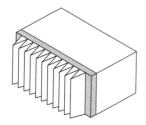

_____ 5. Heat sink mounting method is ___.

_____ 6. Heat sink mounting method is ___.

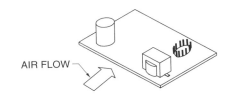

_____ 7. Heat sink mounting method is ___.

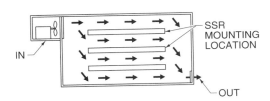

_____ 8. Heat sink mounting method is ___.

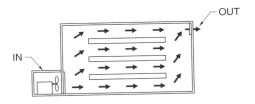

Activity 14-4: Solid-State Relay Installation

Understanding manufacturer specifications is required when selecting, installing, and troubleshooting solid-state relays. Answer the questions using the Manufacturer Data Sheet.

_____ 1. What is the maximum current from a load controlled by the relay?

_____ 2. What is the maximum voltage from a load controlled by the relay?

_____ 3. What is the minimum current from a load controlled by the relay?

_____ 4. What is the minimum voltage from a load controlled by the relay?

MANUFACTURER DATA SHEET

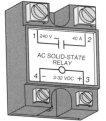

TRIAC AC SSR
- AC solid-state relay
- Zero switching
- Low-cost triac type
- 10 A load current

SSR SPECIFICATIONS	
Input	
Control voltage range	3 – 32 VDC
Pick-up voltage, max.	3 V
Drop-out voltage, min.	1 V
Input impedance	1.5 kΩ
Output	
Load current, max.	10 A
Load current, min.	20 mA
Line voltage range	24 VAC – 280 VAC
Line frequency range	47 – 63 Hz
Thermal Data*	
Operating temp. range	–40 to +100
Storage temp. range	–40 to +100

* in °C

Activity 14-5: Troubleshooting at Motor Starters

When troubleshooting at motor starters, approximate meter readings should be anticipated if the meter readings are going to be used to help determine circuit problems. Determine the expected DMM readings if the circuit is operating properly.

_____ 1. The expected reading of DMM 1 with the motor OFF is ___ VAC.

_____ 2. The expected reading of DMM 1 with the motor ON is ___ VAC.

_____ 3. The expected reading of DMM 2 with the motor OFF is ___ VAC.

_____ 4. The expected reading of DMM 2 with the motor ON is ___ VAC.

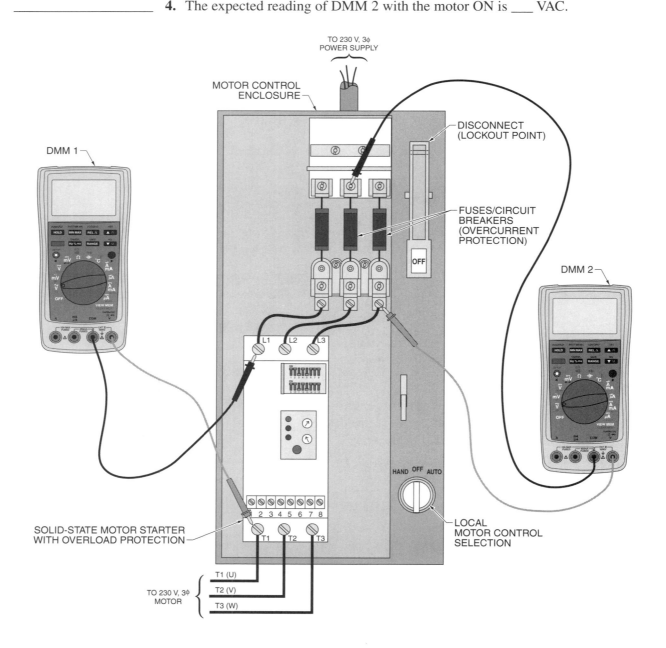

Sensing Devices and Controls

15 Applications

Application 15-1: Proximity Sensors

Sensors

A *proximity sensor* is an electrical switch that does not require physical contact for activation. The three types of proximity sensors are inductive, capacitive, and photoelectric.

Inductive sensors detect metallic objects. Nominal sensing distances range from 0.5 mm to 50 mm. The maximum sensing distance depends on the size of the object to be sensed and the type of metal. For example, iron is sensed at twice the distance of aluminum. Applications for inductive sensors include positioning, fan blade detection, drill bit breakage, and solid-state replacement of mechanical limit switches. **See Inductive Sensor.**

Capacitive sensors detect conductive and nonconductive solid, fluid, or granulated substances. Nominal sensing distances range from 3 mm to 20 mm. The maximum sensing distance depends on the physical and electrical characteristics (dielectric) of the object to be detected. Materials with larger dielectric numbers are easier to detect with a capacitive sensor. Applications for capacitive sensors include sensing the level of products such as sugar, grain, and sand in a container. **See Capacitive Sensor.**

Photoelectric sensors detect most materials and have the greatest sensing distance. Nominal sensing distances range to over 150 cm for sensors that do not use a separate reflector or receiver. Photoelectric sensors that include a separate receiver are available in ranges over

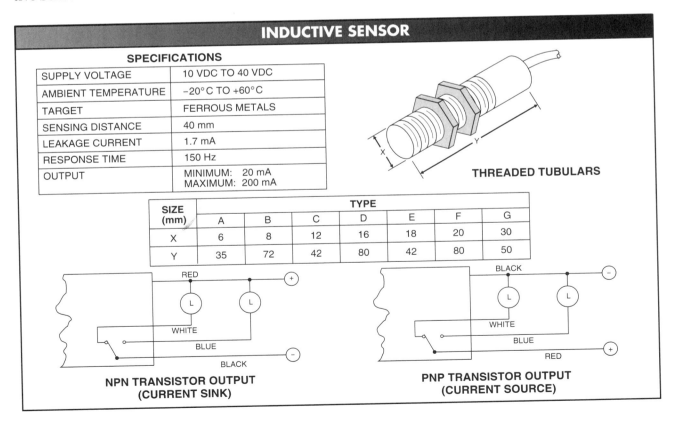

100′. The maximum sensing distance depends on the color and type of surface of the reflecting object. Applications using photoelectric sensors include detecting an object moving along a conveyor system and no-touch detection. when the object to be detected is excessively hot, light, or untouchable. **See Photoelectric Sensor.**

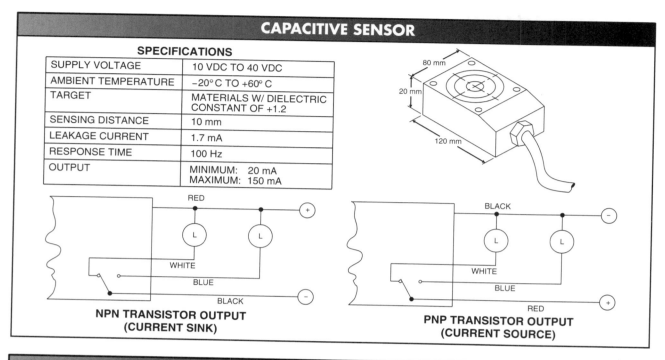

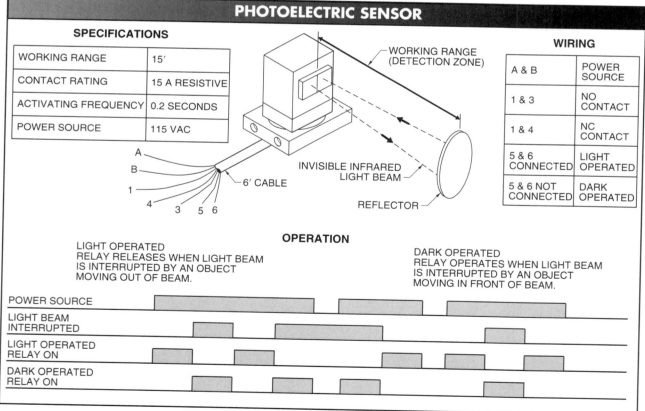

Application 15-2: Proximity Sensor Installation

Sensor Installation

Proximity sensors have a sensing head that produces a radiated sensing field. This sensing field detects the target of the sensor. The sensing field must be kept clear of interference for proper operation. *Interference* is any object other than the object to be detected that is sensed by a sensor. Interference may come from objects close to the sensor or from other sensors. General clearances are required for most proximity sensors.

Flush-Mounted Inductive and Capacitive Proximity Sensors

When flush mounting inductive and capacitive proximity sensors, a distance equal to or greater than twice the diameter of the sensors is required between sensors. If two sensors of different diameter are used, the diameter of the largest sensor is used for installation. **See Flush Mounted.** For example, if two 8 mm inductive proximity sensors are flush mounted, at least 16 mm is required between the sensors.

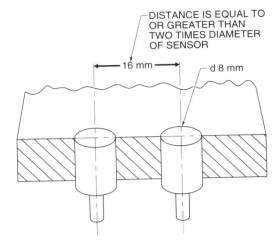

FLUSH MOUNTED

Non-Flush Mounted Inductive and Capacitive Proximity Sensors

When non-flush mounting inductive and capacitive proximity sensors, a distance of three times the diameter of the sensor is required within or next to a material that can be detected. Three times the diameter of the largest sensor is required when inductive and capacitive proximity sensors are installed next to each other. Spacing is measured from center to center of the sensors. When inductive and capacitive proximity sensors are mounted opposite each other, six times the rated sensing distance is required for proper operation. **See Non-Flush Mounted.** For example, if two 16 mm capacitive proximity sensors are non-flush mounted, at least 48 mm is required between the sensors.

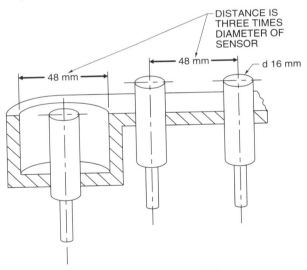

NON-FLUSH MOUNTED

Mounting Photoelectric Sensors

Photoelectric sensors transmit a beam of light. The beam of light detects the presence or absence of an object. Only part of the light beam is effective when detecting the object. The *effective light beam* is the area of light that travels directly from the transmitter to the receiver. If the object does not completely block the effective light beam, the object is not detected. The body of a photoelectric sensor may be cylindrical, rectangular, or slotted.

When mounting photoelectric sensors, the receiver is positioned to receive as much light as possible from the transmitter. Because more light is available at the receiver, greater operating distances are allowed and more power is available for the system to see through dirt in the air and on the transmitter and receiver lenses. The transmitter is mounted on the clean side of the detection zone because light scattered by dirt on the receiver lens affects the system less than light scattered by dirt on the transmitter lens. **See Photoelectric Sensor Positioning.**

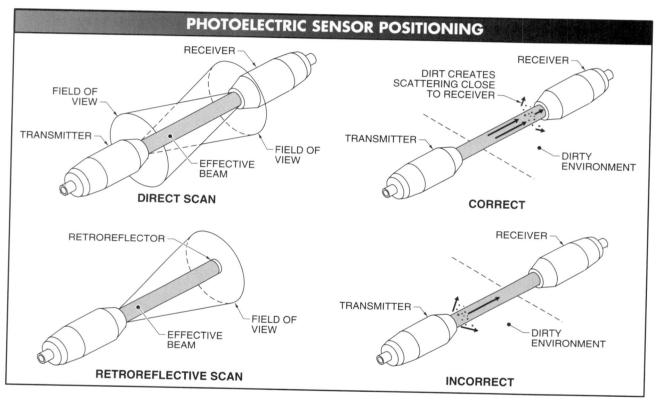

Application 15-3: Determining Activating Frequency

Frequency

Activating frequency is the limit to the number of pulses per second that can be detected by a photoelectric control in a time period. All photoelectric controls have an activating frequency. To determine the required activating frequency of a photoelectric application, apply the following procedure:

1. Determine the maximum speed of the objects to be detected. The speed of the objects is found by measuring the speed of the objects (in ft/min, or in./sec) and converting the speed to sec/in. **See Speed Conversions.** For example, cartons on a conveyor travel at 21′/min. Twenty-one feet per minute equals 0.238 sec/in. (from Speed Conversions table).

2. Determine the dark input signal duration. *Dark input signal duration* is the time period when a photoelectric sensor is dark because the detected object is blocking the light beam. Dark input signal duration is found by multiplying the minimum dimension of the object to be detected (in in.) by the sec/in. value. For example, if a 6″ × 3″ container has a 0.263 sec/in. value, the dark input signal duration is 0.789 sec (3 × 0.263 = 0.789).

3. Determine the light input signal duration. *Light input signal duration* is the time period when a photoelectric sensor is lit because no detectable object is in the light beam. Light input signal duration is found by multiplying the minimum distance between the objects to be detected by the sec/in. value. For example, if the moving objects are spaced between 4″ and 15″ on a conveyor traveling at 0.200 sec/in., the light input signal duration is 0.8 sec (4 × 0.200 = 0.8).

4. Determine activating frequency. Activating frequency of a photoelectric control application is found by adding the dark input signal duration to the light input signal duration. This value is compared to the manufacturer stated value. The activating frequency of a photoelectric control must be less than the activating frequency required for the application.

SPEED CONVERSIONS

Ft/min	In./min	Ft/sec	In./sec	Sec/in.
1	12	0.017	0.2	5
3	36	0.050	0.6	1.666
5	60	0.083	1.0	1.000
7	84	0.116	1.4	0.714
9	108	0.150	1.8	0.555
11	132	0.183	2.2	0.435
13	156	0.216	2.6	0.385
15	180	0.249	3.0	0.333
17	204	0.282	3.4	0.294
19	228	0.315	3.8	0.263
21	252	0.349	4.2	0.238
23	276	0.382	4.6	0.2172
25	300	0.415	5	0.200
40	480	0.664	8	0.125
60	720	0.996	12	0.0833
80	960	1.328	16	0.0625
100	1200	1.66	20	0.050
200	2400	3.32	40	0.025
250	3000	4.15	50	0.020
300	3600	4.98	60	0.016
400	4800	6.64	80	0.012
500	6000	8.30	100	0.010
700	8400	11.62	140	0.007
1250	15,000	20.75	250	0.004
2500	30,000	45.5	500	0.002

Example: Finding Activating Frequency
The minimum distance between objects in an application is 5″. The dimensions of the objects are 2″ × 2″. The maximum speed of travel is 40′/min. Find the activating frequency.

1. Determine maximum speed of travel in sec/in.

   Maximum speed of travel is 0.125 sec/in. (from Speed Conversions table).

2. Determine dark input signal duration.

   Dark input signal duration is 0.25 sec (2 × 0.125 = 0.25).

3. Determine light input signal duration.

   Light input signal duration is 0.625 sec (5 × 0.125 = 0.625).

4. Determine activating frequency.

   Activating frequency is 0.875 sec (0.25 + 0.625 = 0.875). To work properly, the photoelectric control must have an activating frequency of 0.875 sec or less.

Application 15-4: Applying Photoelectric Sensors

Photoelectric Sensors

Photoelectric sensors are used in many applications because they detect the presence or absence of any object without touching the object. Because of this no-touch feature, proximity sensors can be used in applications where mechanical limit switches cannot. The sensors can detect light objects, heavy objects, or untouchable objects. Untouchable objects include hot objects and freshly painted objects. **See Proximity Sensor Applications.**

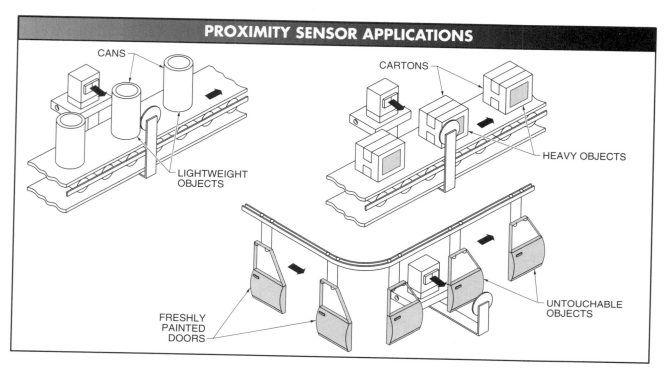

Application 15-5: Troubleshooting Photoelectric Sensors

Photoelectric Switch Operation

Troubleshooting circuits that include photoelectric switches requires an understanding of how the switch contacts operate in the circuit. If the photoelectric switch is set for dark operation, the switch's contacts change state (normally open close and normally closed open) any time there is an object in front of the eye (blocking light beam). If the photoelectric switch is set for light operation, the switch's contacts change state (normally open close and normally closed open) any time there is not an object in front of the eye (light beam not blocked).

In many applications, the photoelectric switch contacts are used to control a timer so that there is a time delay produced in the circuit. The timer contacts are then used to energize or de-energize a load. Thus, when troubleshooting an application that uses a timer with a photoelectric sensor, both the operation of the photoelectric sensor (light operated or dark operated) and the type of timer (on-delay, off-delay, one-shot, or recycle) must be understood. Photoelectric switches are used to detect the presence of vehicles at tollgates, parking areas, and truck docks. The photoelectric switch signals are used as inputs to timers, relays, counters, PLCs, and motor control circuits.

Sensing Devices and Controls

15 Activities

Name _____ Date _____

Activity 15-1: Proximity Sensors

In the bottling application, a capacitive proximity switch is used to detect bottle flow along the system. Identify the color of each wire of the proximity switch using the Capacitive Sensor on page 220.

_____ 1. Wire 1 color is ___.

_____ 2. Wire 2 color is ___.

_____ 3. Wire 3 color is ___.

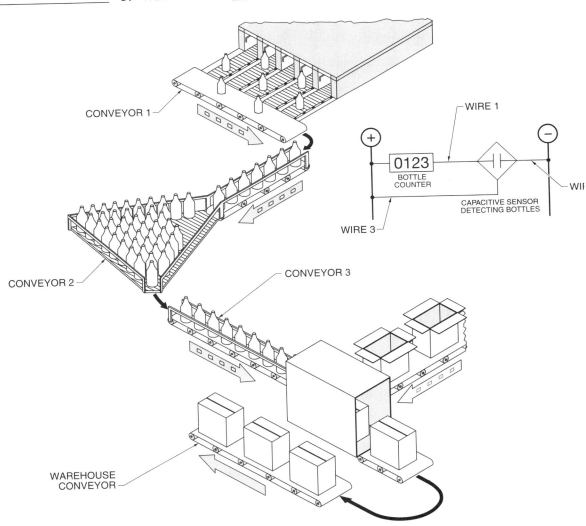

226 ELECTRICAL MOTOR CONTROLS *for Integrated Systems* APPLICATIONS MANUAL

In the bottling application, a photoelectric switch is used to detect cartons moving along the system. Identify the number/letter of each wire of the photoelectric switch using the Photoelectric Sensor on page 220.

_____ **4.** Wire 1 number/letter is ___.

_____ **5.** Wire 2 number/letter is ___.

_____ **6.** Wire 3 number/letter is ___.

_____ **7.** Wire 4 number/letter is ___.

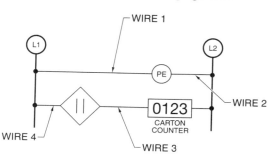

Activity 15-2: Proximity Sensor Installation

Determine the minimum distance required between the sensors for proper operation.

_____ **1.** The minimum distance required between the sensors for proper operation is ___ mm.

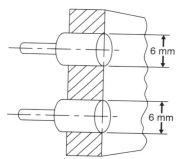

_____ **2.** The minimum distance required between the sensor and the surrounding material for proper operation is ___ mm.

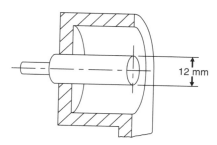

_____ 3. The minimum distance required between the sensors and the surrounding material for proper operation is ___ mm.

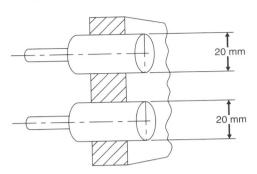

Activity 15-3: Determining Activating Frequency

Answer the questions using the Speed Conversions chart on page 223.
A photoelectric sensor detects 2" × 2" objects that are 5" apart and travel at 60'/min.

_____ 1. The dark input signal duration is ___ sec.

_____ 2. The light input signal duration is ___ sec.

_____ 3. The activating frequency of the application is ___ sec.

A photoelectric sensor detects 0.5" square objects that are 2" apart and travel at 200'/min.

_____ 4. The dark input signal duration is ___ sec.

_____ 5. The light input signal duration is ___ sec.

_____ 6. The activating frequency of the application is ___ sec.

A photoelectric sensor detects 2" square objects that are 3" apart and travel at 1250'/min.

_____ 7. The dark input signal duration is ___ sec.

_____ 8. The light input signal duration is ___ sec.

_____ 9. The activating frequency of the application is ___ sec.

A photoelectric sensor detects 0.25" × 0.25" objects that are 0.25" apart and travel at 15'/min.

_____ 10. The dark input signal duration is ___ sec.

_____ 11. The light input signal duration is ___ sec.

_____ 12. The activating frequency of the application is ___ sec.

Activity 15-4: Applying Photoelectric Sensors

Answer the following questions using the Photoelectric Sensor Data Sheet.

_____ 1. A photoelectric sensor is used in an application to detect when a box is in place. When the box is detected, the photoelectric sensor normally open contacts (pins 1 and 3) are to close in the control circuit. Which photoelectric sensor wiring diagram example (1A, 1B, 2A, or 2B) applies to this application?

_____ 2. A photoelectric sensor is used in an application to detect when an unpainted aluminum roll runs out of material or breaks. When the material is no longer detected (is out or broken), the photoelectric sensor normally closed contacts (pins 1 and 4) reclose in the control circuit to sound an alarm. Which photoelectric sensor wiring diagram example (1A, 1B, 2A, or 2B) applies to this application?

PHOTOELECTRIC SENSOR DATA SHEET

- Relay for photosensor with modulated infrared light
- Built-in power supply for transmitter/receiver
- For separate transmitters and receivers activating distance: 1 m–100 m
- For combined transmitters and reflector activating distance: 1 m–10 m
- Transmitter and receiver connections are short-circuit protected
- 10 A SPDT output relay
- LED indication: relay on

WIRING DIAGRAMS

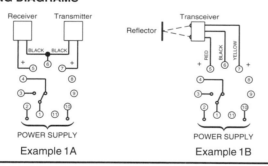
Example 1A   Example 1B

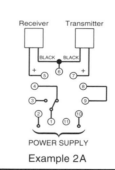
Example 2A

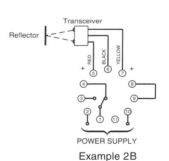
Example 2B

MODE OF OPERATION

Relay is used in conjunction with separate, infrared modulated transmitters, and receivers or transceiver with a reflector.
Detection by reflection is carried out either by using a reflector or various reflective materials such as plastics, metal, or glass.

Example 1 (LIGHT OPERATED)
Relay releases when light beam is interrupted or in case of power failure. The relay operates when the receiver short-circuits.

Example 2 (DARK OPERATED)
Relay operates when light beam is interrupted. Relay releases when the receiver short-circuits. Interconnect pins 8 and 9 directly on the base.

OPERATIONAL DIAGRAM

| Supply voltage |
| Light beam interrupted |
| Ex 1 Relay ON (light operated) |
| Ex 2 Relay ON (dark operated) |

Activity 15-5: Troubleshooting Photoelectric Sensors

When troubleshooting conveyor systems containing photoelectric controls, approximate meter readings should be anticipated if the meter readings are going to be used to help determine circuit problems. Determine the expected DMM readings if the circuit is operating properly.

_____ 1. The expected reading of DMM 1 with the conveyor ON without boxes of goods on the conveyor belt is ___ VAC.

_____ 2. The expected reading of DMM 1 with the conveyor ON with boxes of goods on the conveyor belt backed up to the photo 2 position is ___ VAC.

_____ 3. The expected reading of DMM 2 with the conveyor ON without boxes of goods on the conveyor belt is ___ VAC.

_____ 4. The expected reading of DMM 2 with the conveyor ON with boxes of goods on the conveyor belt backed up to the photo 2 position is ___ VAC.

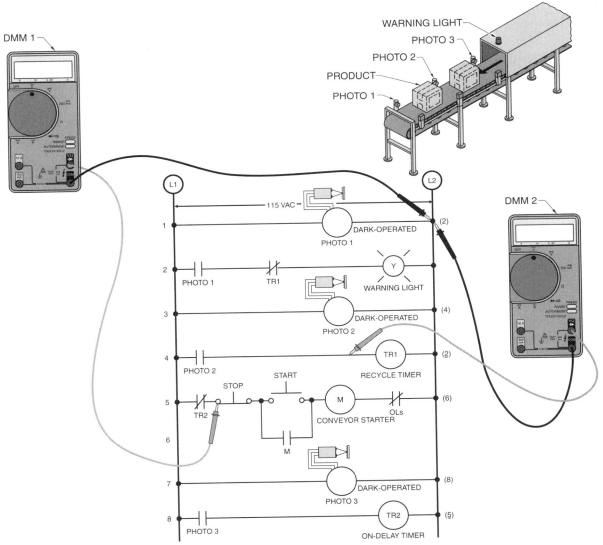

When troubleshooting truck loading bay systems containing photoelectric controls, approximate meter readings should be anticipated if the meter readings are going to be used to help determine circuit problems. Determine the expected DMM readings if the circuit is operating properly.

_____ 5. The expected reading of DMM 1 without a truck in the loading bay is ___ VAC.

_____ 6. The expected reading of DMM 1 with a truck in the loading bay is ___ VAC.

_____ 7. The expected reading of DMM 2 without a truck in the loading bay is ___ VAC.

_____ 8. The expected reading of DMM 2 with a truck in the loading bay is ___ VAC.

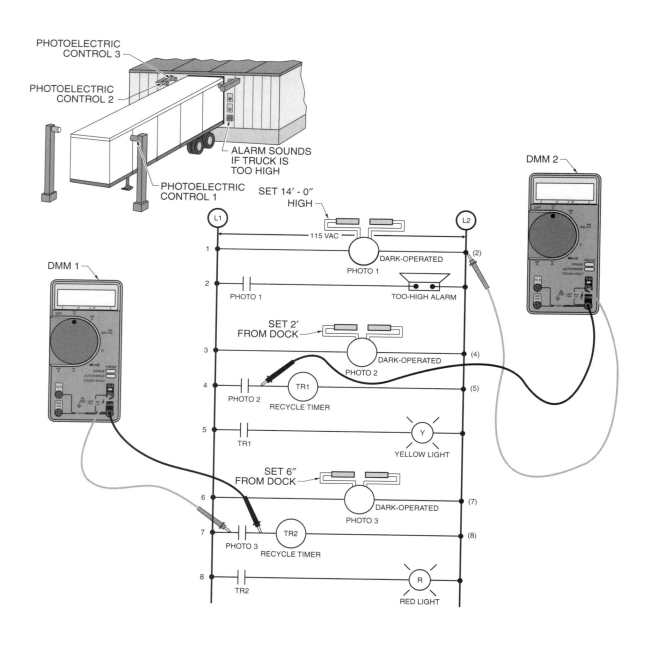

Programmable Controllers

16 Applications

Application 16-1: Programmable Controller Input and Output Identification

Programmable Controllers

A programmable controller is used to control machine or process operations with a stored program that controls the outputs based on the status of the inputs. No wiring change of the inputs and outputs is required when the machine function or process is changed. The logic of the circuit is changed through the programming terminal. If a control circuit uses a programmable controller, the inputs and outputs must be identified when designing, wiring, or troubleshooting a system. **See Programmable Controller System.**

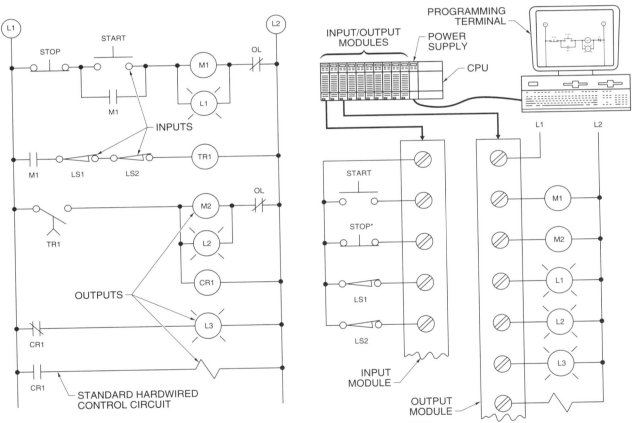

* NOTE: 1. HARDWIRED NORMALLY OPEN PUSHBUTTON IS PROGRAMMED AS NORMALLY CLOSED
2. HARDWIRED NORMALLY CLOSED PUSHBUTTON IS PROGRAMMED AS NORMALLY OPEN

PROGRAMMABLE CONTROLLER SYSTEM

Input Section

The input section of a programmable controller receives information from pushbuttons, temperature switches, pressure switches, overload contacts, and other manual, mechanical, or automatic inputs.

The inputs connected to a programmable controller are classified as digital or analog. A *digital input* is an input that has only two positions, ON and OFF. Digital inputs include pushbuttons, switches, relay contacts, and mechanical limit switches. An *analog input* is an input that changes continuously over a range. Analog inputs include variable-voltage inputs, variable-current inputs, and variable-resistance inputs. Analog inputs include potentiometers, rheostats, and temperature, pressure, and humidity transducers. The input section receives incoming signals and converts them to low-power digital signals that are sent to the processor section.

Output Section

The output section of a programmable controller delivers the output voltage to control alarms, lights, solenoids, motor starters, and other devices that produce work in a system. Like inputs, outputs connected to a programmable controller are digital or analog.

Digital outputs include ON or OFF devices such as motor starters, alarms, solenoids, and contactors. Analog outputs include variable-voltage and variable-current outputs such as analog meters, motors, and dimming lights.

If the output of a programmable controller draws more power than the output section can handle, an interface is used. An *interface* is a control device that allows a small current to control a large current. For example, a motor starter connected directly to the output of a programmable controller can be used to control a large motor.

Application 16-2: Programmable Controller Input and Output Connections

Connecting Inputs and Outputs

When connecting an input to a programmable controller, one side of the input is connected to an assigned input terminal and the other side is connected to a common terminal. The assigned input terminal is usually marked IN 1, IN 2, IN 12, IN 125, etc. The common terminal is usually marked COM or C. When connecting an output to a programmable controller, one side of the output is connected to an assigned output terminal and the other side is connected to a common power line. The assigned output terminal is usually marked OUT 1, OUT 2, OUT 12, OUT 125, etc. The power line connections are usually marked L1/L2, +/–, +V/–V, VAC IN, or VDC IN. **See Input and Output Connections.**

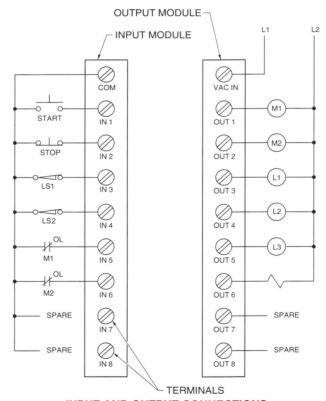

INPUT AND OUTPUT CONNECTIONS

Application 16-3: Alarm Output Connection

Low Battery Alarm

Programmable controllers include alarm output contacts that are activated if the reserve battery is low or other problems occur in the programmable controller. If the reserve battery is low and a power failure occurs, the program can be lost. When power returns, the programmable controller does not operate properly until it is reprogrammed. This can cause problems in a system. **See Programmable Controller Alarm.**

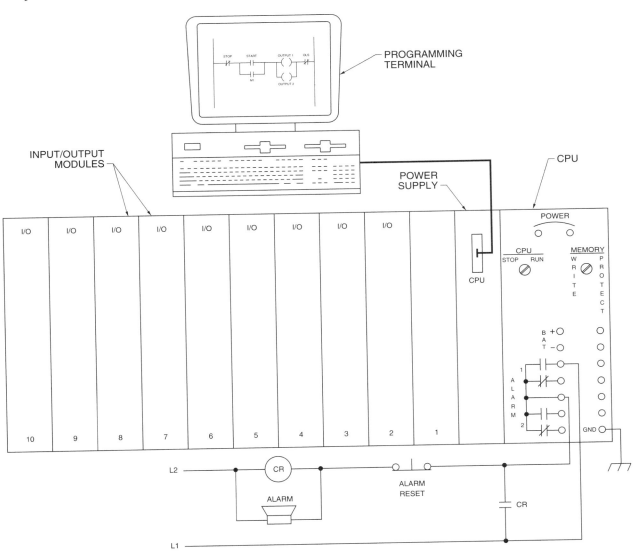

PROGRAMMABLE CONTROLLER ALARM

Application 16-4: Developing PLC Programmable Circuits

Programmable PLC Circuits

In order to program a circuit on a PLC, the circuit must first be drawn in PLC format. When converting a standard line diagram to a PLC diagram, all inputs are placed on the left side of the circuit between the left rung and the output, including the overload (OL) contacts, and all outputs are placed on the right side of the circuit. Inputs can be placed in series, parallel, or in series/parallel combinations. The same inputs can be programmed at multiple locations in the circuit and programmed as normally open or normally closed. Any PLC output (motor starter, lamp, timer, alarm, etc.) can have programmed normally open or normally closed contacts. **See Equivalent Hardwired and PLC Circuits.**

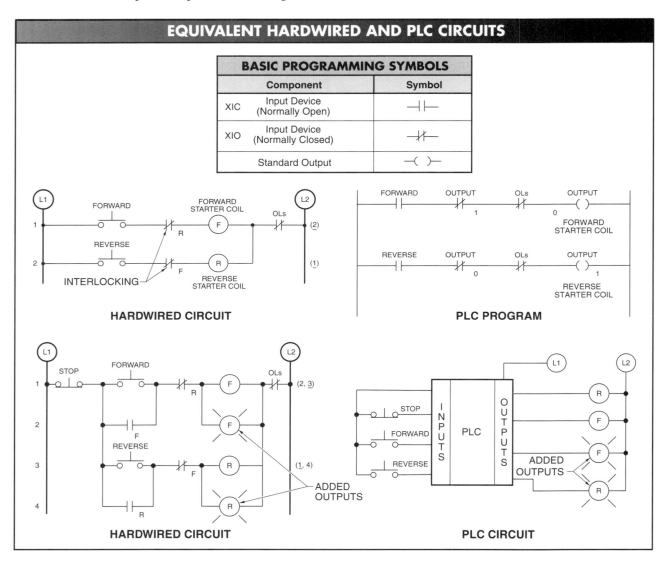

Application 16-5: Troubleshooting PLC Inputs and Outputs

Troubleshooting Input Modules

Signals and information are sent to a programmable controller using input devices such as pushbuttons, limit switches, level switches, and pressure switches. The input devices are connected to the input module of the programmable controller. The controller does not receive the proper information if the input device or input module is not operating correctly. **See Troubleshooting Input Modules.**

To troubleshoot an input module, apply the following procedure:

1. Measure the supply voltage at the input module to ensure that there is power supplied to the input device(s). Test the main power supply of the controller when there is no power.
2. Measure the voltage from the control switch. Connect a DMM directly to the same terminal screw to which the input device is connected.

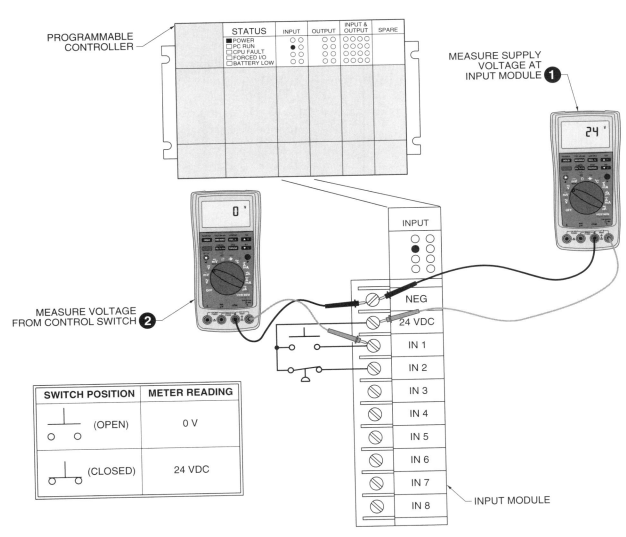

TROUBLESHOOTING INPUT MODULES

The DMM should read the supply voltage when the control switch is closed. The DMM should read the full supply voltage when the control device uses mechanical contacts. The DMM should read nearly the full supply voltage when the control device is solid-state. Full supply voltage is not read because 0.5 V to 6 V is dropped across the solid-state control device. The DMM should read zero or little voltage when the control switch is open.

Troubleshooting Output Modules

A programmable controller turns the output devices (loads) in the circuit ON and OFF according to the program. The output devices are connected to the output module of the programmable controller. No work is produced in the circuit when the output module or output devices are not operating correctly. When an output device does not operate, the problem may lie in the output module, output device, or controller. **See Troubleshooting Output Modules.**

To troubleshoot an output module, apply the following procedure:

1. Measure the supply voltage at the output module to ensure that there is power supplied to the output devices. Test the main power supply of the controller when there is no power.
2. Measure the voltage delivered from the output module. Connect a DMM directly to the same terminal screw to which the output device is connected.

The DMM should read the supply voltage when the program energizes the output device. The DMM should read full supply voltage when the output module uses mechanical contacts. The DMM should read almost full supply voltage when the output module uses a solid-state switch. Full voltage is not read because 0.5 V to 6 V is dropped across the solid-state switch. The DMM should read zero or little voltage when the program de-energizes the output device.

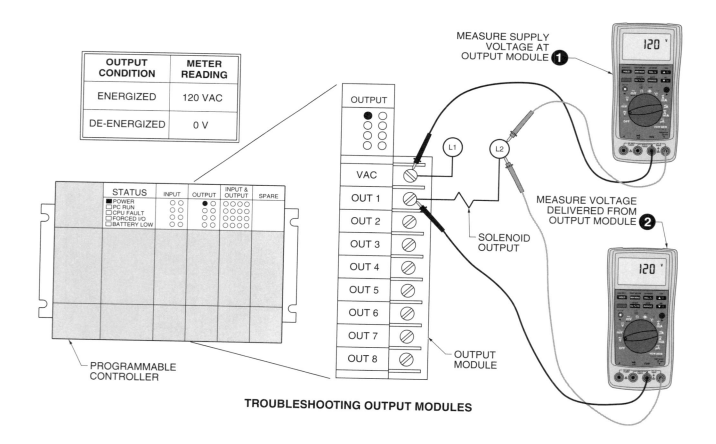

TROUBLESHOOTING OUTPUT MODULES

Programmable Controllers

16 Activities

Name _____ Date _____

Activity 16-1: Programmable Controller Input and Output Identification

1. Connect the circuit inputs and outputs to the PLC using the overhead door circuit line diagram. Identify the order (input 1, input 2, etc.) of the inputs as they are used in the line diagram as read from left to right and top to bottom. Identify the order (output 1, output 2, etc.) of the outputs as they are used in the line diagram as read from left to right and top to bottom.

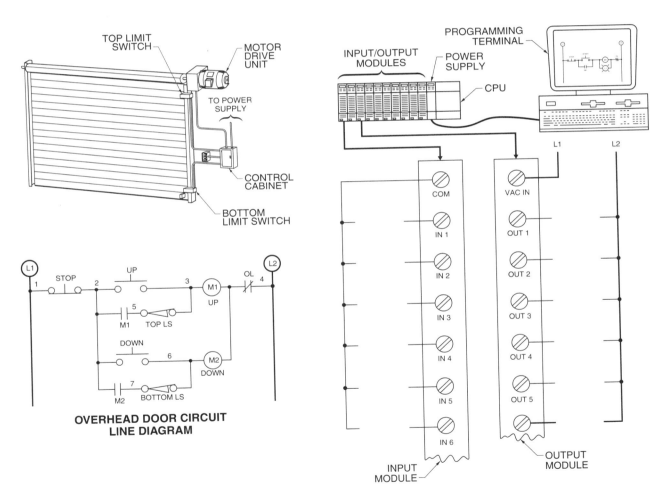

237

Activity 16-2: Programmable Controller Input and Output Connection

Electric cylinders are output devices that use electricity (not air or oil) to produce a linear mechanical force on a cylinder rod. Electric cylinders are used in applications that require variable position control or where pressurized fluid is not available. Electric cylinders are driven by AC or DC motors. Any type of switch can control an electric cylinder motor. In the damper control application, pushbuttons are used to control damper motor 1, a three-position selector switch controls damper motor 2, and a joystick controls damper motor 3.

1. Connect the circuit inputs to the input section of the PLC so that only the normally open contacts of the pushbuttons are used. The advantage of a PLC is that the equivalent normally closed contacts of the pushbuttons can be programmed when programming the PLC circuit, thus eliminating the need for any more than one set of normally open contacts wired to the PLC. Connect the three-position selector switch so that PLC input 3 is activated when the selector switch is in the left position and input 4 is activated when the selector switch is in the right position. Connect the four-position joystick so that input 5 is activated when the joystick is placed in the up position and input 6 is activated when the joystick is placed in the down position. Connect the open and closed starter coils of damper motors 1, 2, and 3 to the PLC.

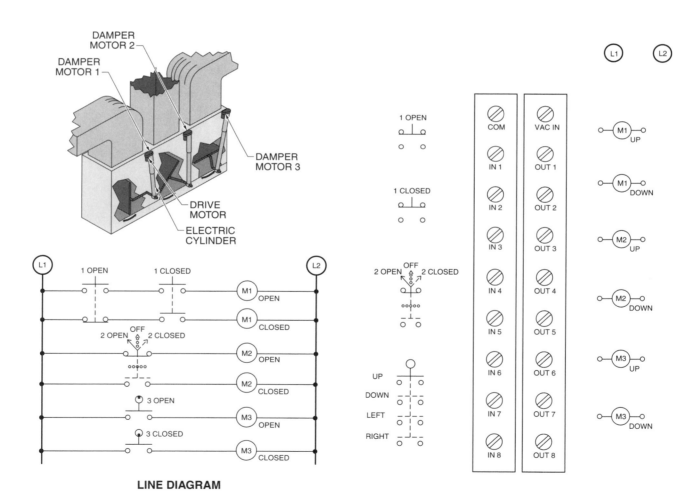

LINE DIAGRAM

Activity 16-3: Alarm Output Connection

An alarm circuit sounds if there is a problem with a PLC. The alarm may be in a location that can be heard or may not be heard in a noisy environment. Indicating lamps can be added at the operator station to give a visual indication of the alarm circuit operating condition.

1. Connect the red lamp at the operator station into the alarm circuit so that it turns ON if there is a problem. Connect the green lamp so that the lamp is ON if there is no problem and turns OFF if a problem occurs.

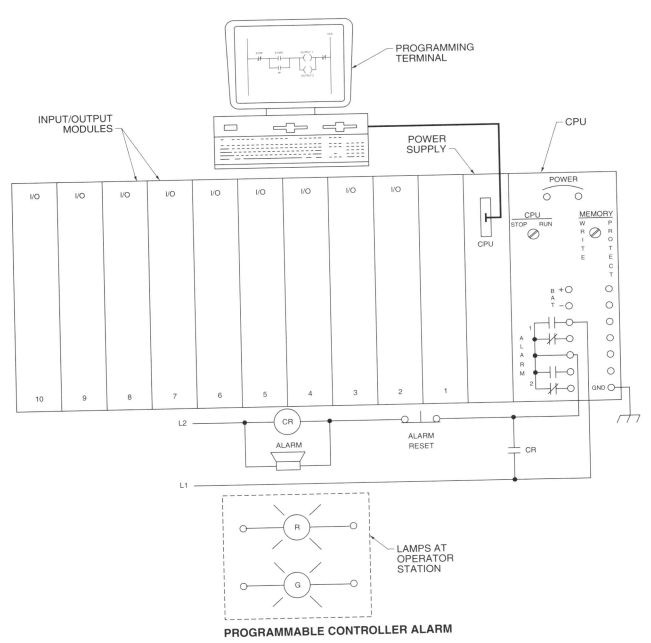

PROGRAMMABLE CONTROLLER ALARM

Activity 16-4: Developing PLC Programmable Circuits

Convert each standard line diagram into a PLC diagram. Label all PLC symbols to match the given circuit. Use the standard PLC symbols for normally open, normally closed, and outputs when drawing the circuit. Mark each input and output (stop, start, level switch, motor starter, etc.).

1.

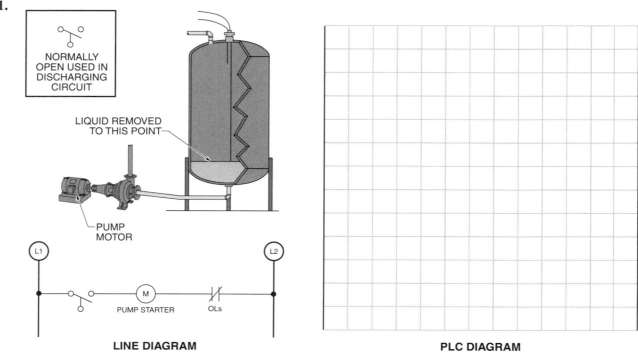

LINE DIAGRAM

PLC DIAGRAM

2.

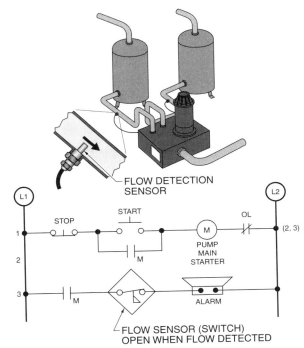

LINE DIAGRAM

PLC DIAGRAM

3.

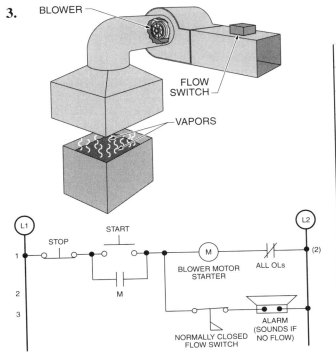

LINE DIAGRAM

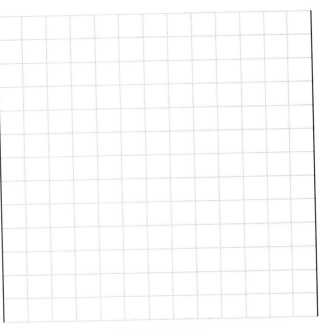

PLC DIAGRAM

4.

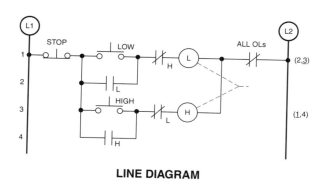

LINE DIAGRAM

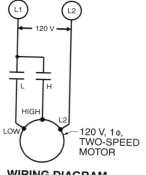

WIRING DIAGRAM

PLC DIAGRAM

5.

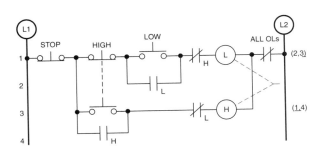
LINE DIAGRAM

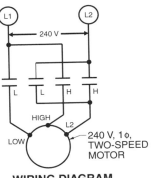
WIRING DIAGRAM

PLC DIAGRAM

Activity 16-5: Troubleshooting PLC Inputs and Outputs

When troubleshooting programmable controllers, approximate meter readings should be anticipated if the meter readings are going to be used to help determine circuit problems. Determine the expected DMM readings if the circuit is operating properly.

_____ 1. The expected reading of DMM 1 with M1 OFF is ___ VAC.

_____ 2. The expected reading of DMM 1 with M1 ON is ___ VAC.

_____ 3. The expected reading of DMM 2 with LS2 closed is ___ VAC.

_____ 4. The expected reading of DMM 2 with LS2 open is ___ VAC.

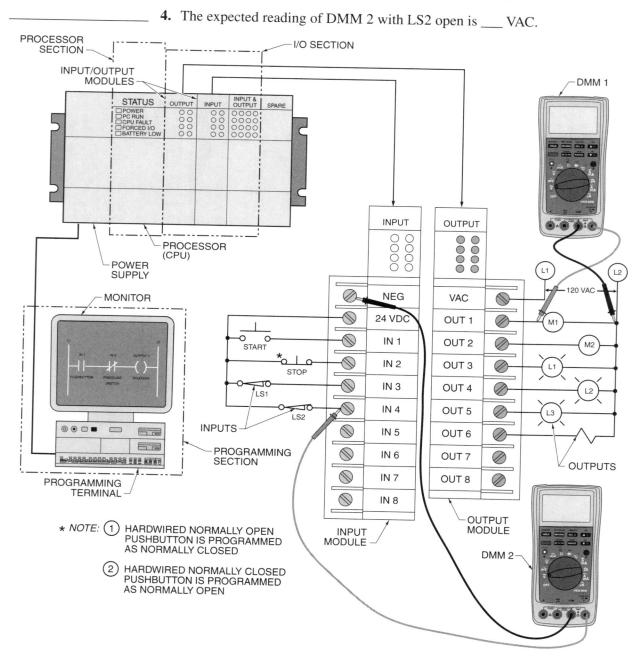

Reduced-Voltage Starting Applications 17

Application 17-1: Primary Resistor Reduced-Voltage Starting

Primary Resistor Starting

Primary resistor reduced-voltage starting is a motor starting method that has resistors connected in series with a motor to reduce the voltage applied to the motor when starting. After a time period, full line voltage is applied to the motor. Primary resistor reduced-voltage starting provides a smooth acceleration through a simple, economical circuit. Primary resistor reduced-voltage starting applications include gear or belt drives, especially in industries in which paper, clothing, and other delicate fabrics are produced, where the sudden application of the full-voltage torque must be avoided. **See Primary Resistor Reduced-Voltage Starting.**

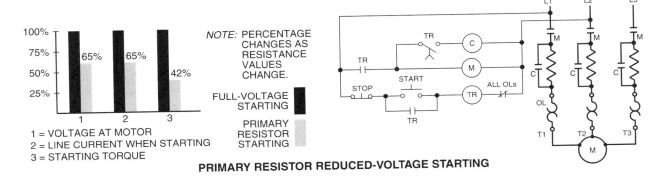

PRIMARY RESISTOR REDUCED-VOLTAGE STARTING

Application 17-2: Part-Winding Reduced-Voltage Starting

Part-Winding Starting

Part-winding reduced-voltage starting is a motor starting method that applies current to one-half of the motor windings when starting. After a time period, the starter applies current to all motor windings. Part-winding reduced-voltage starting is economical because it requires fewer components and less space than other reduced-voltage starting methods. Part-winding reduced-voltage starting applications include low-inertia loads such as commercial air conditioning compressors, pumps, fans, and blowers, or locations where local power companies place limitations on the amount of inrush current. Part-winding reduced-voltage starting requires a nine-lead, 3φ motor. **See Part-Winding Reduced-Voltage Starting.**

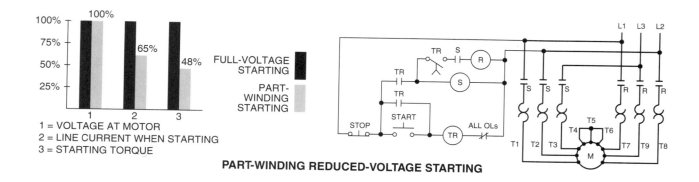

PART-WINDING REDUCED-VOLTAGE STARTING

Application 17-3: Autotransformer Reduced-Voltage Starting

Autotransformer Starting

Autotransformer reduced-voltage starting is a motor starting method that reduces the applied motor voltage to 50%, 65%, or 80% of the line voltage when starting. This is accomplished by placing a transformer coil in series with the motor for a time period. After the time period, the motor is connected to full line voltage. Autotransformer reduced-voltage starting is used for starting blower, compressor, conveyor, and pump motors over 10 HP. Autotransformer starting is one of the most effective methods of reduced-voltage starting.

Autotransformer reduced-voltage starting provides the highest possible starting torque per ampere of line current. However, because autotransformers are required, installation cost is higher than other reduced-voltage starting methods. Autotransformer reduced-voltage starting is used with any 3ϕ motor. **See Autotransformer Reduced-Voltage Starting.**

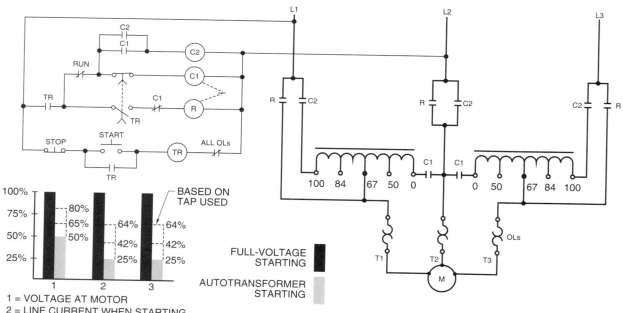

AUTOTRANSFORMER REDUCED-VOLTAGE STARTING

Application 17-4: Wye/Delta Reduced-Voltage Starting

Wye/Delta Starting

Wye/delta reduced-voltage starting is a motor starting method that has the motor connected as a wye motor when starting. This arrangement reduces the coil voltage of the motor to about 58% of line voltage. After a time period, the motor is connected as a delta motor. Wye/delta reduced-voltage starting is used where the power supply is inadequate to provide full starting current without an objectionable voltage drop or where low starting torque is required. Wye/delta reduced-voltage starting applications include fans, compressors, and conveyors that have long acceleration times or frequent starts. Wye/delta reduced-voltage starting is only used with six-lead, 3ϕ motors. **See Wye/Delta Reduced-Voltage Starting.**

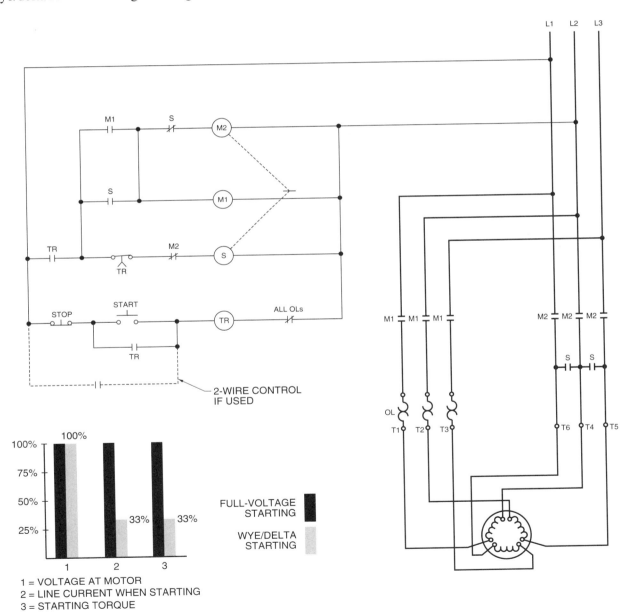

1 = VOLTAGE AT MOTOR
2 = LINE CURRENT WHEN STARTING
3 = STARTING TORQUE

WYE/DELTA REDUCED-VOLTAGE STARTING

Application 17-5: Closed Transition Reduced-Voltage Starting

Closed Transition Starting

Closed transition reduced-voltage starting is a method of motor starting in which the motor is not disconnected from the power source during the transition from start to run. There is a brief time period between the point when a motor is disconnected from the reduced voltage and the point when it is reconnected to full line voltage in autotransformer or wye/delta reduced-voltage starting. This open transition from start to run causes a flow of high transient current when the motor is reconnected to full line voltage. To eliminate this problem, closed transition starting is added. **See Closed Transition Reduced-Voltage Starting.**

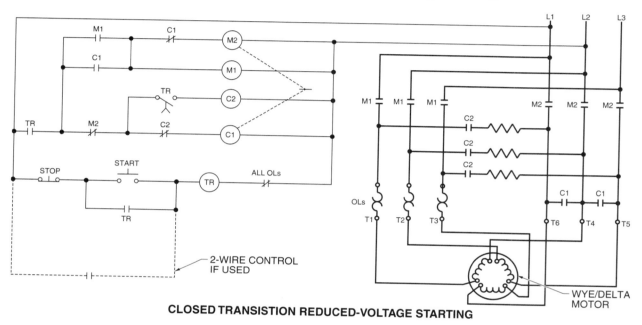

CLOSED TRANSISTION REDUCED-VOLTAGE STARTING

Application 17-6: Troubleshooting Reduced-Voltage Starting Circuits

Troubleshooting Reduced-Voltage Starting

Troubleshooting reduced-voltage starting circuits follows the same basic troubleshooting procedures used when troubleshooting full voltage starting circuits. When troubleshooting reduced-voltage starting circuits, a DMM is used to test the voltage of the power circuit. If the power circuit voltage is not correct, the problem is upstream in the power circuit (main disconnect, disconnect fuses, etc.). If the power circuit voltage is correct, the voltage of the control circuit is checked. If the voltage of the control circuit is not correct, the control transformer is checked. If the control circuit voltage is correct, voltage measurements are taken throughout the control circuit, testing the voltage into and out of each component until the problem is found.

Reduced-Voltage Starting

17 Activities

Name _____   Date _____

Activity 17-1: Primary Resistor Reduced-Voltage Starting

The voltage of a control circuit should be 120 V or less. A control transformer is used to step down the high power circuit voltage (230 V, 460 V, etc.) to a low control circuit voltage (115 V, 24 V, 12 V, etc.). A dual-voltage primary control transformer is used because it can be wired for either of two different input voltages. When wiring a control transformer, the transformer windings are placed in series for high voltage and in parallel for low voltage.

1. Wire the control transformer into the primary resistor reduced-voltage starter circuit.

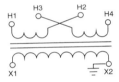

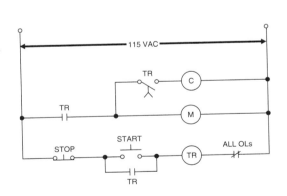

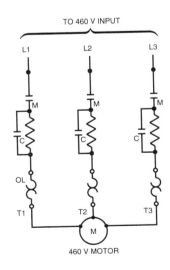

249

250 ELECTRICAL MOTOR CONTROLS *for Integrated Systems* APPLICATIONS MANUAL

Activity 17-2: Part-Winding Reduced-Voltage Starting

The voltage of a control circuit should be 120 V or less. A control transformer is used to step down the high power circuit voltage (230 V, 460 V, etc.) to a low control circuit voltage (115 V, 24 V, 12 V, etc.). A dual-voltage primary control transformer is used because it can be wired for either of two different input voltages. When wiring a control transformer, the transformer windings are placed in series for high voltage and in parallel for low voltage. To protect the control circuit, a fuse or circuit breaker can be added to the control circuit. Protection devices such as fuses and circuit breakers are added to the hot (ungrounded) side of the circuit.

1. Wire the control transformer and fuse into the part-winding reduced-voltage starting circuit.

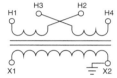

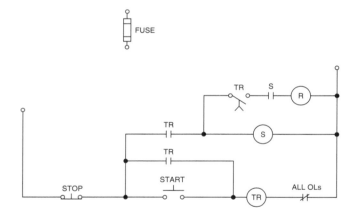

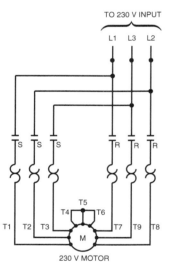

Activity 17-3: Autotransformer Reduced-Voltage Starting

Relays, contactors, motor starters, and timers use contacts in control circuits to produce the required circuit logic.

1. For each relay, starter, and timer, add the cross-reference numbering system to the right of the device. Also connect the control transformer and fuse into the circuit.

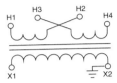

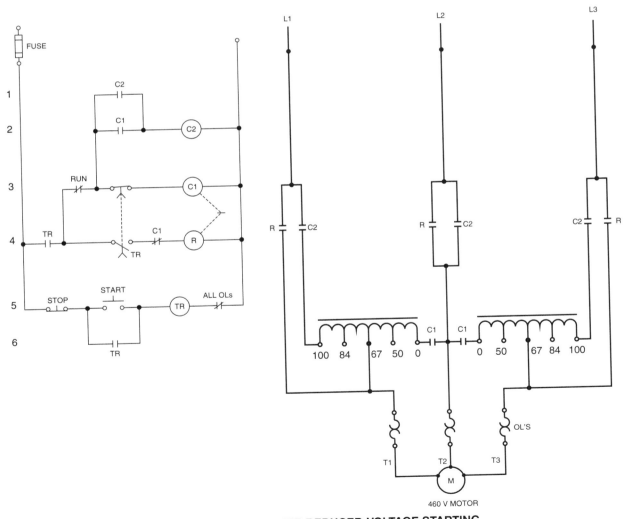

AUTOTRANSFORMER REDUCED-VOLTAGE STARTING

Activity 17-4: Wye/Delta Reduced-Voltage Starting

Circuits are often designed for a standard start/stop pushbutton control. However, the circuit may actually have to operate using a temperature switch, pressure switch, or some other two-wire control device.

1. Modify the wye/delta reduced-voltage circuit by circling the part of the control circuit that would be removed and marking it "remove" and adding a pressure switch to control the circuit in its place.

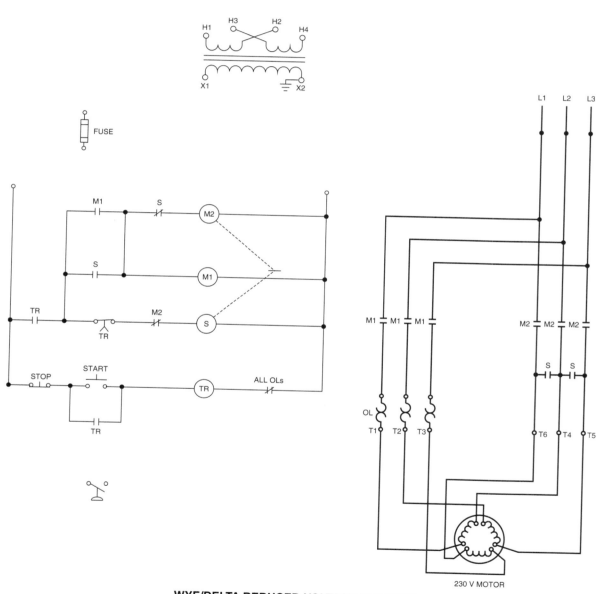

WYE/DELTA REDUCED-VOLTAGE STARTING

Activity 17-5: Closed Transition Reduced-Voltage Starting

Each wire in a control circuit is assigned a wire reference (terminal) point on the line diagram. Each reference point is assigned a reference number.

1. Connect the control transformer and fuse into the circuit. Assign wire reference numbers to each wire in the control circuit. Assign X1 the number 1 and X2 the number 2. Continue the numbering system by assigning the wire coming out of the fuse (and to the stop pushbutton) the number 3 and number all other wires in the circuit from left-to-right and top-to-bottom. Circle each wire reference (terminal) number.

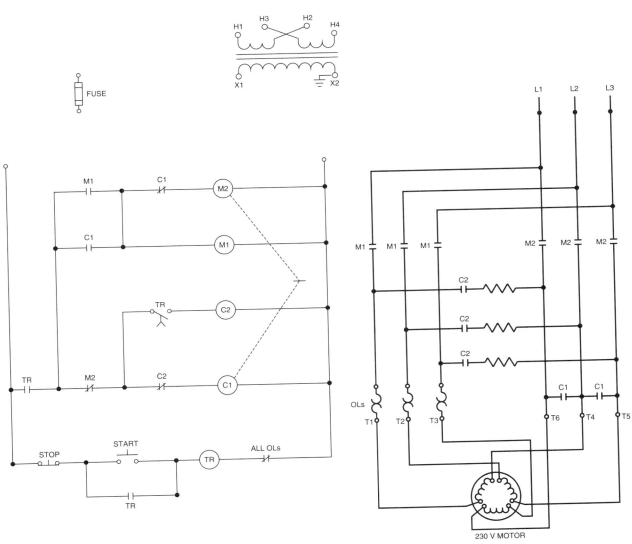

CLOSED TRANSISTION REDUCED-VOLTAGE STARTING

Activity 17-6: Troubleshooting Reduced-Voltage Starting Circuits

Answer the questions using Primary Resistor Reduced-Voltage Starting Circuit. Note: The motor operates when the start pushbutton is pressed, but does not accelerate to full power. The machine operator reports that there is a burning smell coming from the control cabinet. Once the motor is started, the resistors get hot after several minutes. A voltage reading taken at the motor terminals indicates lower voltage than voltage readings taken at the incoming power lines.

_____ 1. A(n) ___ is the most likely cause of the malfunction.

_____ 2. Could the problem be in the pushbutton station?

_____ 3. Could the problem be in the motor starter?

_____ 4. Could the problem be in the contactor?

_____ 5. Could the problem be in the resistors?

_____ 6. Could the problem be in the timer?

_____ 7. Could the problem be in the overload contact?

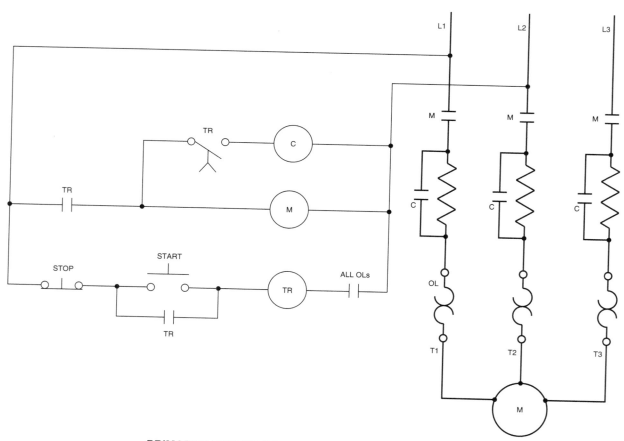

PRIMARY RESISTOR REDUCED-VOLTAGE STARTING CIRCUIT

When troubleshooting reduced-voltage starting circuits, approximate meter readings should be anticipated if the meter readings are going to be used to help determine circuit problems. Determine the expected DMM readings if the circuit is operating properly.

_____ 8. The expected reading of DMM 1 with the start pushbutton pressed and with TR1 not timed out is ___ VAC.

_____ 9. The expected reading of DMM 1 with the start pushbutton pressed and with TR1 timed out is ___ VAC.

_____ 10. The expected reading of DMM 2 with the motor turned ON is ___ VAC.

_____ 11. The expected reading of DMM 2 with the motor turned OFF is ___ VAC.

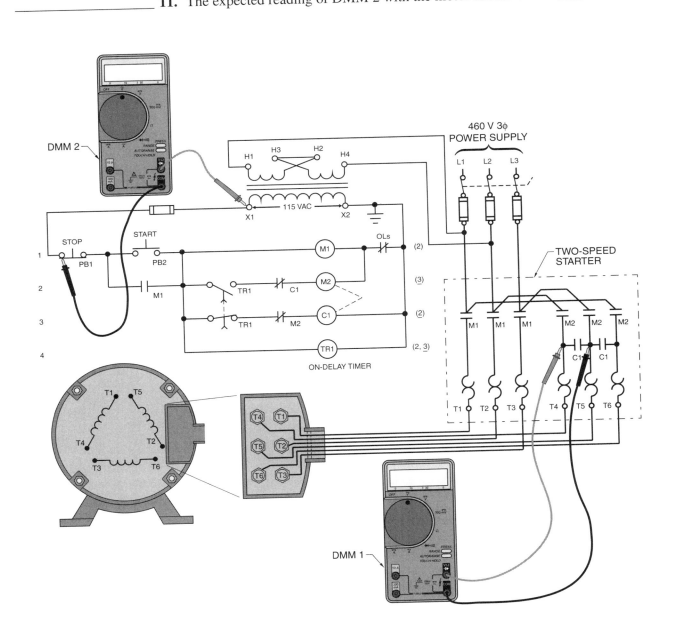

Accelerating and Decelerating Methods

Applications 18

Application 18-1: One-Direction Motor Plugging

Motor Plugging—One-Direction

Plugging is a method of motor braking in which motor connections are reversed so the motor develops a countertorque that acts as a braking force. Plugging a motor to a rapid stop is done with a plugging switch. A plugging switch prevents the motor from reversing direction when the motor comes to a stop. A plugging switch automatically interrupts the reversing braking power as a motor approaches 0 rpm. The speed at which plugging switch contacts operate is adjusted to avoid coasting or reverse rotation of a motor. Plugging is used in emergency shutdowns for safety reasons. Plugging is also used to protect machines and machine tools. When using plugging as a stopping method, the following factors should be considered:

- The driven machine, belts, chains, or couplings must be designed to withstand the forces created by plugging.
- When plugging a motor, approximately six to eight times the normal full-load current of the motor is drawn. The power supply and control equipment must be sized to withstand this overload.
- The motor may not be sized to withstand frequent plugging. A larger motor or a motor with a higher duty rating may be required. **See One-Direction Motor Plugging.**

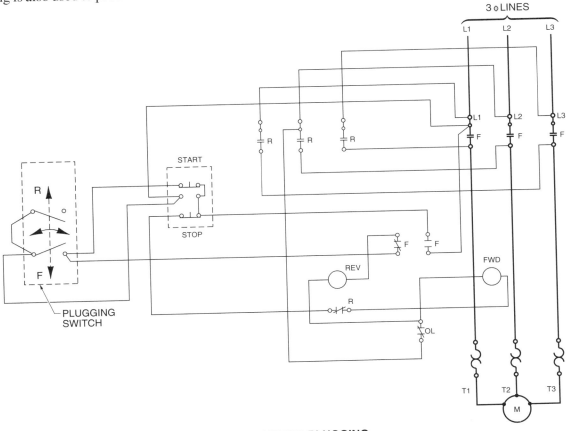

ONE-DIRECTION MOTOR PLUGGING

Application 18-2: Two-Direction Motor Plugging

Motor Plugging—Two-Direction

Plugging produces a very rapid stop of a motor. Plugging can be produced in one motor direction or in both directions. For this reason, most plugging switches are available with a normally open contact that closes in the forward motor direction and a separate normally open contact that closes in the reverse motor direction. When plugging a motor in only one direction, only one set of the normally open plugging contacts is used. When plugging a motor in two directions, both sets of normally open plugging contacts are used. **See Two-Direction Motor Plugging.**

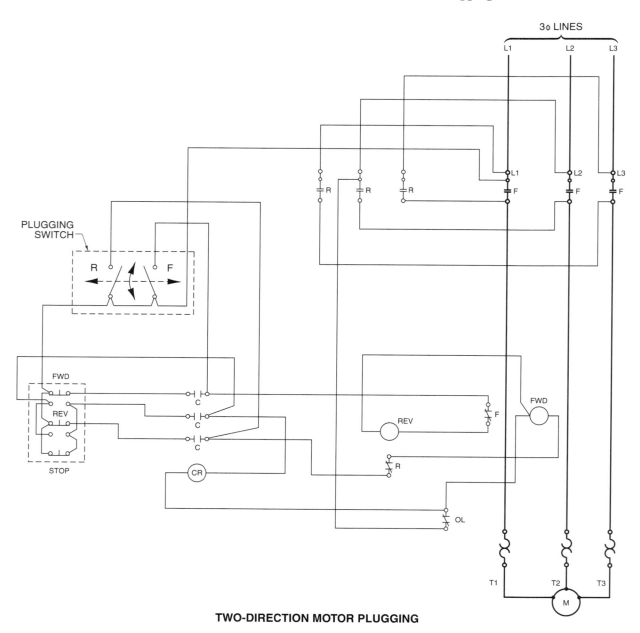

TWO-DIRECTION MOTOR PLUGGING

Application 18-3: Two-Speed Separate Winding Motors

Separate Winding Motors

The speed of an AC motor is determined by the frequency of the power supply or the number of individual poles of the motor. The individual poles are determined by how the motor windings are connected. The speed of a motor decreases as the number of poles increases, and the speed of a motor increases as the number of poles decreases. **See Motor Speed.** To change the speed of a motor, the motor must have separate windings. Each winding has a different number of individual poles. When power is applied to the different windings, the motor speed changes. In a two-speed motor circuit, the motor can be started in either high or low speed.

Two-speed motors are wired the same basic way as any other motor. Basic steps are followed regardless of whether the motor is a single-voltage motor, dual-voltage motor, reversible motor, two-speed or three-speed motor, etc. When wiring any motor, look at the nameplate of the motor for written information and a wiring diagram. The written information states the motor's required voltage and operating speed. The wiring diagram shows how the motor is wired for a low voltage and high voltage on dual-voltage motors. The wiring diagram also shows how the motor is wired for low speed and high speed on dual-speed motors. How the motor is reversed is also listed as part of the information included with the wiring diagram (usually below the wiring diagram).

With all power OFF, the motor wiring diagram is used to connect the motor to the motor side of the starter. Next, with all power OFF, the power lines are connected to the power side of the starter. After the circuit is connected and power applied, the motor is tested at low speed. Only if the motor operates properly at low speed should the motor be tested at high speed.

MOTOR SPEED*

Poles	Synchronous Speed	With a 4% Slip	Actual Speed
2	3600	144	3456
4	1800	72	1728
6	1200	48	1152
8	900	36	864

* in rpm

Application 18-4: Two-Speed Consequent Pole Motors

Consequent Pole Motors

Motor speed can be changed with consequent pole motors. A *consequent pole motor* is a motor that has one winding connected and reconnected so that it has half or twice the original number of poles. As the number of poles on the winding changes, the speed of the motor changes. **See Two-Speed Consequent Pole Motor.**

The number of poles changes because of the way the current from the power lines (L1, L2, or L3) travels through the motor windings. Since each motor winding is divided into two equal parts, the current can be made to travel through the windings connected in series or parallel with each other.

STARTER CONNECTIONS				
SPEED	SUPPLY LINES		OPEN	TOGETHER
	L1	L2	L3		
LOW	T1	T2	T3	NONE	T4, T5, T6
HIGH	T6	T4	T5	T1, T2, T3	NONE

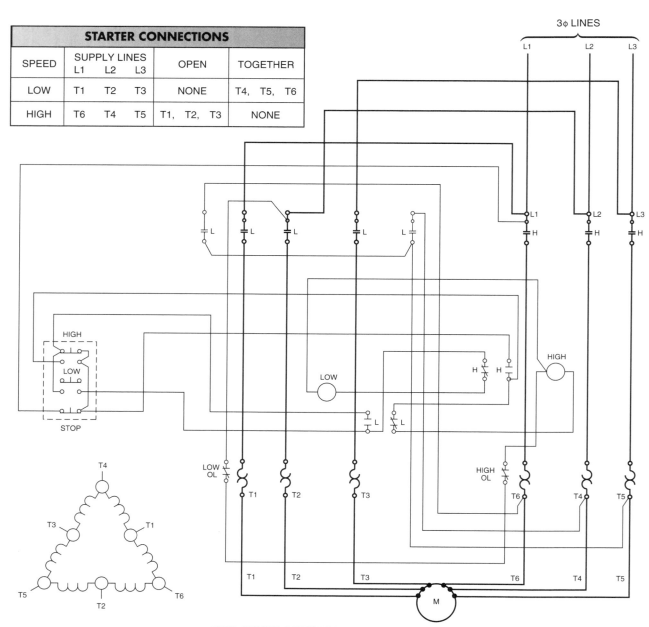

TWO-SPEED CONSEQUENT POLE MOTOR

Application 18-5: Motor Torque and Horsepower

Motor Full-Load Torque

Full-load torque is the torque required to produce the rated power at a motor's full speed.

Motor torque is found by applying the following formula:

$$T = \frac{HP \times 5252}{rpm}$$

where
T = torque (in ft-lb)
HP = horsepower
5252 = constant $\left(\dfrac{33{,}000}{\pi \times 2}\right)$
rpm = revolutions per minute

A motor that is fully loaded produces full-load torque. A motor that is underloaded produces less than full-load torque. A motor that is overloaded must produce more than full-load torque to keep the load operating at rated speed. The higher the produced full-load torque, the higher the motor current draw and internal heating of the motor.

Motor Horsepower

Horsepower is used to measure the energy produced by an electric motor while doing work. Motor horsepower is found by applying the following formula:

$$HP = \frac{rpm \times T}{5252}$$

where
HP = horsepower
rpm = revolutions per minute
T = torque (in ft-lb)
5252 = constant $\left(\dfrac{33{,}000}{\pi \times 2}\right)$

> **Example: Calculating Motor Full-Load Torque**
> What is the full-load torque of a 2 HP motor operating at 1200 rpm?
>
> $T = \dfrac{HP \times 5252}{rpm}$
>
> $T = \dfrac{2 \times 5252}{1200}$
>
> $T = \dfrac{10{,}504}{1200}$
>
> $T = \mathbf{8.75\ lb\text{-}ft}$

> **Example: Calculating Motor Horsepower**
> What is the horsepower of a 1200 rpm motor with a full-load torque of 8.75 lb-ft?
>
> $HP = \dfrac{rpm \times T}{5252}$
>
> $HP = \dfrac{1200 \times 8.75}{5252}$
>
> $HP = \dfrac{10{,}500}{5252}$
>
> $HP = \mathbf{2\ HP}$

Application 18-6: Troubleshooting Two-Speed Circuits

Two-Speed Separate Winding Circuits

When troubleshooting a two-speed circuit, the power and control circuits must be checked. If the control circuit is not working correctly, the power circuit will not work. Thus, troubleshooting can start with a few quick checks in the control and power circuits.

When troubleshooting a two-speed circuit, a digital multimeter set to measure voltage is placed across the low-speed magnetic motor starter coil. The motor is set to operate at the low speed (low-speed pushbutton pressed). The control circuit is operating properly if the full control voltage is present across the low-speed motor starter.

If the low-speed motor starter is operating properly, the voltage at the low-speed motor windings is measured. If no voltage is present at the low-speed motor windings, there is a problem with the motor starter power contacts. If the correct voltage is present and the motor is not operating, the motor windings are probably open (bad motor).

If the motor operates properly at low speed, a DMM set to measure voltage is placed across the high-speed magnetic motor starter coil. The motor is set to operate at high speed (high-speed pushbutton pressed). The control circuit is operating properly if the full control voltage is present across the high-speed motor starter.

If the high-speed motor starter is operating properly, the voltage at the high-speed motor windings is measured. If there is no voltage, there is a problem with the motor starter power contacts. If the voltage is correct and the motor is not operating, the motor windings are probably open (bad motor).

Application 18-7: Troubleshooting Two-Direction Plugging Circuits

Two-Direction Motor Plugging Circuits

When an electrical system or circuit is not operating correctly, troubleshooting is done to determine the problem. In order to reduce troubleshooting time, the electrician must first start testing the most likely component that has failed. This requires knowledge of what each component in the circuit does. For example, if no part of a circuit is operating, the problem is usually with the fuses or circuit breakers. Fuses and circuit breakers remove power from a circuit to prevent the circuit from operating in an overloaded condition. Open fuses in the power circuit prevent both the power circuit and control circuit from operating. Open fuses in the control circuit prevent only the components in the control circuit (starters, lamps, etc.) from operating. Fuses are checked if the control circuit is not operating.

Control transformers take high voltage for the power circuit and step down the voltage for the control circuit. If no components in the control circuit are operating, the voltage into (primary side) and out of (secondary side) the transformer is measured.

Most control switches (pushbuttons, limit switches, plugging switches, etc.) perform a specific function. When a control switch is the problem, usually only part of the circuit is affected. If only part of a circuit is operating properly, the control switches in the part of the circuit that is not operating properly are checked.

Application 18-8: Troubleshooting Electronic Braking Circuits

Electronic Braking Circuits

When troubleshooting an all AC circuit, a DMM can be set on AC voltage and measurements taken throughout the circuit. Likewise, when troubleshooting a DC circuit, a DMM can be set on DC voltage and measurements taken throughout the circuit. However, when a circuit includes both AC and DC, DMM settings (AC or DC) and the point in the circuit in which the DMM leads are connected must be correct for the type of voltage present. If a voltage level is unknown or does not seem right, start by setting the DMM to measure DC voltage. The measurement is taken and the DMM leads reversed and the measurement taken again. The voltage is DC if both readings are at the same level and one reading displays a negative sign. If the voltage is not DC, take the readings again with the DMM set on AC.

The DC power usually comes from a rectifier circuit connected to the AC circuit. If there appears to be a DC problem, measure the voltage out of the rectifier (DC measurement) and into the rectifier (AC measurement). If the rectifier is operating properly and the problem seems to be with the DC circuit, start troubleshooting the rest of the DC circuit one component at a time. If the DC circuit seems to be operating properly, troubleshoot the AC circuit one component at a time.

Accelerating and Decelerating Methods

Activities 18

Name _____ Date _____

Activity 18-1: One-Direction Motor Plugging

Plugging requires that an interlocked forward and reversing motor starter be used. The starters must be interlocked to prevent the motor from being connected to operate in forward and reverse directions at the same time. The exact model (type) forward/reversing starter ordered for the plugging application depends on the motor type and control circuit used. Answer the questions using the One-Direction Motor Plugging circuit on page 257 and the Forward/Reversing Interlocked Starter Types.

_____ 1. A Type ___ (1, 2, 3, etc.) forward/reversing starter is required for the application in which the starter has the minimum number of required contacts for the given circuit.

_____ 2. List all the forward/reversing starter types that could be used for the application, even if they exceed the number of required contacts.

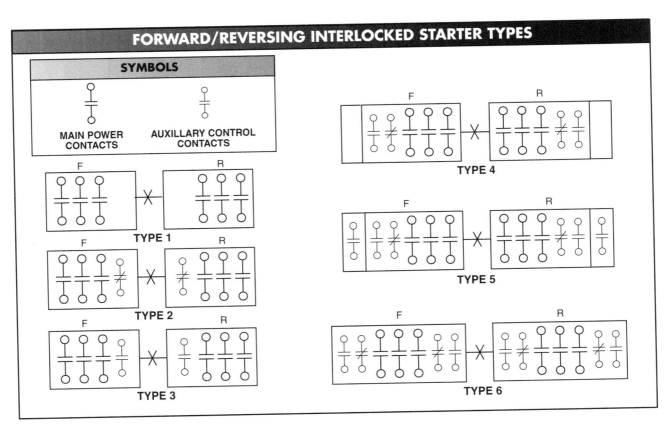

263

Activity 18-2: Two-Direction Motor Plugging

Plugging a motor in two directions requires a relay with multiple contacts in the control circuit. Answer the questions using the Two-Direction Motor Plugging circuit on page 258 and the Control Relay Types.

_____ 1. A Type ___ (A, B, C, etc.) relay is required for the application in which the relay has the minimum number of required contacts for the given circuit.

_____ 2. List all of the control relay types that could be used for the application, even if they exceed the number of required contacts.

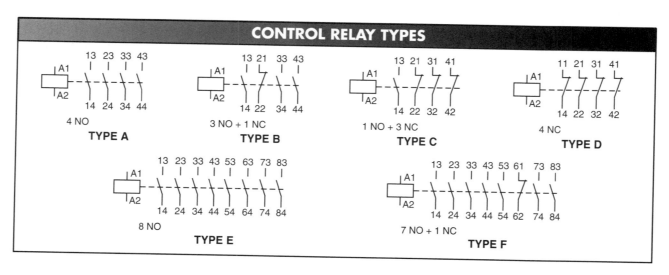

Activity 18-3: Two-Speed Separate Winding Motors

1. Connect the incoming power lines (L1, L2, and L3) to the low-speed (L) and high-speed (H) input side of the power contacts and connect the motor to the output side of the power contacts using the two-speed motor wiring diagram.

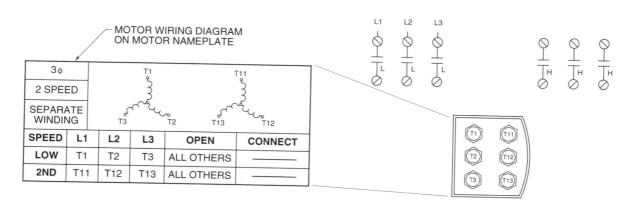

Activity 18-4: Two-Speed Consequent Pole Motors

Answer the questions using the Two-Speed Consequent Pole Motor circuit on page 260 and the Low/High Speed Interlocked Starter Types.

_____ 1. A Type ___ (1, 2, 3, etc.) low/high speed starter is required for the application in which the starter has the minimum number of required contacts for the given circuit.

_____ 2. List all of the low/high speed starter types that could be used for the application, even if they exceed the number of required contacts.

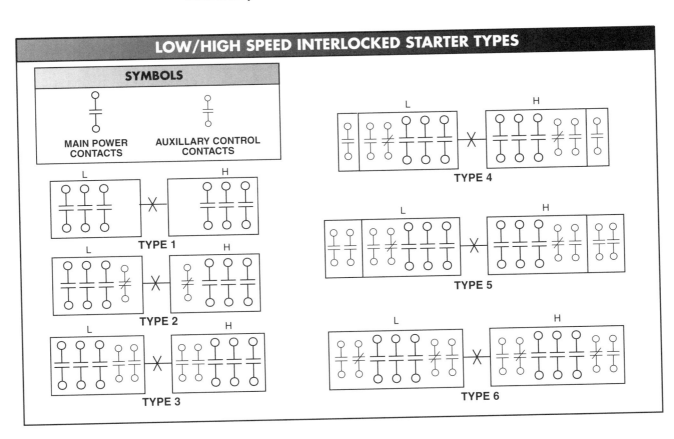

Activity 18-5: Motor Torque and Horsepower

Answer the questions using the motor torque formula, Three-Phase Motor, and Electric Motor Drive tables on page 267.

_____ 1. A catalog number M-2A motor produces ___ ft-lb of torque when connected to 230 V.

_____ 2. A catalog number M-2A motor produces ___ ft-lb of torque when connected to 460 V.

_____ 3. A catalog number M-2B motor produces ___ ft-lb of torque when connected to 230 V.

_____ 4. A catalog number M-2B motor produces ___ ft-lb of torque when connected to 460 V.

_____ 5. If a catalog number M-2B motor is used in a 460 V system that includes a motor drive, a catalog number ___ motor drive would be required.

_____ 6. The total quoted price of both the motor and drive is $ ___.

Answer the questions using the motor torque formula, Three-Phase Motor, and Electric Motor Drive tables on page 267.

_____ 7. A customer requires a 230 V motor that can theoretically produce 15 ft-lb of torque at 1725 rpm. If the customer requests a motor that is rated 35% higher to ensure proper operation over time, a catalog number ___ motor would be quoted.

_____ 8. The cost of the motor quoted to the customer is $ ___.

_____ 9. If the customer wanted a quote for the next larger size motor in case of an application change, the quote for additional cost would be $ ___.

_____ 10. If the customer wanted a quote on a motor drive for the motor quoted in question 7, the price quote would be $ ___.

_____ 11. If the customer wanted to change the quote in question 7 for a 3450 rpm motor, a catalog number ___ motor would be quoted.

_____ 12. The cost of the 3450 rpm motor quoted to the customer is $ ___.

An application requires a 230 VAC motor operating at 1725 rpm to produce a full-load torque of 10 lb-in.

_____ 13. A catalog number ___ motor that is at least 25% overrated would be ordered for the application.

_____ 14. A catalog number ___ drive would be ordered for the motor size and voltage required for the application.

THREE-PHASE MOTOR

HP	RPM (60 Hz)	Voltage	Catalog Number	List Price*
1	3450	230/460 – 60 Hz	M-1A	415
1	1725	230/460 – 60 Hz	M-1B	400
1½	3450	230/460 – 60 Hz	M-2A	450
1½	1725	230/460 – 60 Hz	M-2B	425
2	3450	230/460 – 60 Hz	M-3A	475
2	1725	230/460 – 60 Hz	M-3B	450
3	3450	230/460 – 60 Hz	M-4A	525
3	1725	230/460 – 60 Hz	M-4B	500
5	3450	230/460 – 60 Hz	M-5A	675
5	1725	230/460 – 60 Hz	M-5B	640
7½	3450	230/460 – 60 Hz	M-6A	775
7½	1725	230/460 – 60 Hz	M-6B	725
10	3450	230/460 – 60 Hz	M-7A	900
10	1725	230/460 – 60 Hz	M-7B	850
15	3450	230/460 – 60 Hz	M-8A	1400
15	1725	230/460 – 60 Hz	M-8B	1350

* In $

ELECTRIC MOTOR DRIVE

HP	Input Voltage	Output Current Continuous*	Output Current Peak*	Catalog Number	List Price†
1	230	2	8	D-LV-1	850
1	460	4	4	D-HV-1	900
2	230	7	14	D-LV-2	925
2	460	4	8	D-HV-2	975
3	230	10	20	D-LV-3	1150
3	460	5	10	D-HV-3	1275
5	230	16	32	D-LV-4	1300
5	460	8	16	D-HV-4	1360
7½	230	22	44	D-LV-5	1750
7½	460	11	22	D-HV-5	1840
10	230	28	56	D-LV-6	2400
10	460	14	28	D-HV-6	2525
15	230	42	84	D-LV-7	300
15	460	21	42	D-HV-7	3200
20	230	55	100	D-LV-8	3700
20	460	27	54	D-HV-8	3850

* In A
† In $

An application requires a 460 VAC motor operating at 1725 rpm to produce a full-load torque of 10 lb-in.

_____ **15.** A catalog number ___ motor that is at least 25% overrated would be ordered for the application.

_____ **16.** A catalog number ___ drive would be ordered for the motor size and voltage required for the application.

Activity 18-6: Troubleshooting Two-Speed Circuits

Identify the malfunctioning component using the Two-Speed Separate Winding Circuit and the meter readings. Note: The machine operator reports that the motor operates at the low speed but not at the high speed. DMM 1 reads 440 V when the low pushbutton is pressed. DMM 2 reads 440 V when the low pushbutton is pressed. DMM 3 reads 440 V when the high pushbutton is pressed. DMM 4 reads 440 V when the high pushbutton is pressed.

_____ **1.** The component that has failed is the ___.

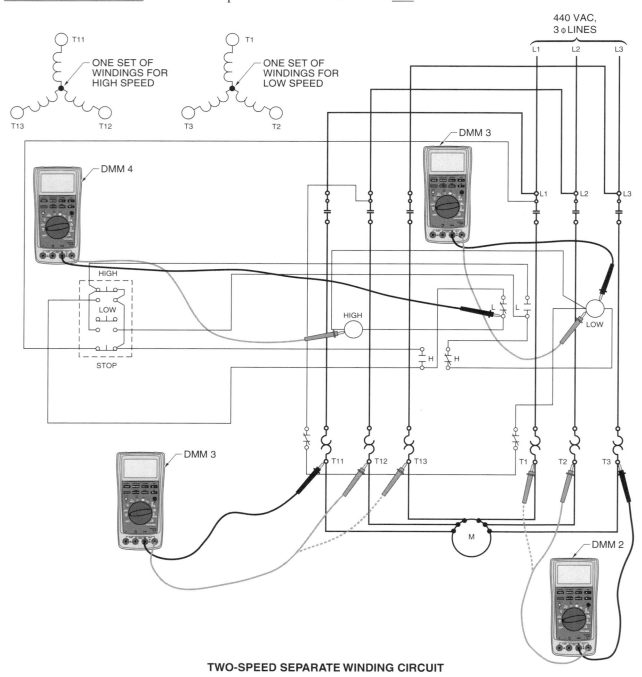

TWO-SPEED SEPARATE WINDING CIRCUIT

Activity 18-7: Troubleshooting Two-Direction Plugging Circuits

1. Connect DMM 1 to monitor the malfunctioning component when the motor is turned OFF in the forward direction. Connect DMM 2 to monitor the malfunctioning component when the motor is turned OFF in the reverse direction. *Note:* The machine operator reports that the motor does not come to a rapid stop in either direction.

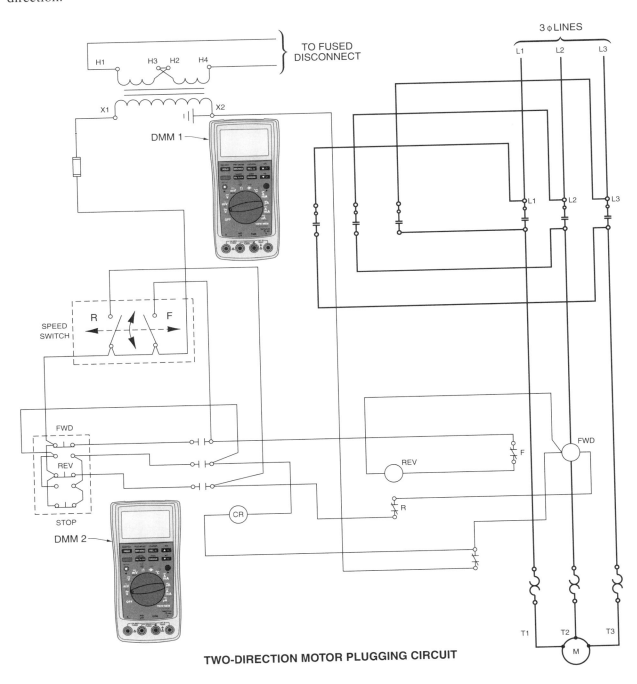

TWO-DIRECTION MOTOR PLUGGING CIRCUIT

Activity 18-8: Troubleshooting Electronic Braking Circuits

When troubleshooting electronic braking circuits, approximate meter readings should be anticipated if the meter readings are going to be used to help determine circuit problems. Determine the expected DMM readings if the circuit is operating properly.

_____ 1. The expected reading of DMM 1 if the stop pushbutton is pressed while the motor is running is ___ VDC.

_____ 2. The expected reading of DMM 1 with the timer fully timed out is ___ VDC.

_____ 3. The expected reading of DMM 2 with the motor running is ___ VAC.

_____ 4. The expected reading of DMM 2 with the motor OFF is ___ VAC.

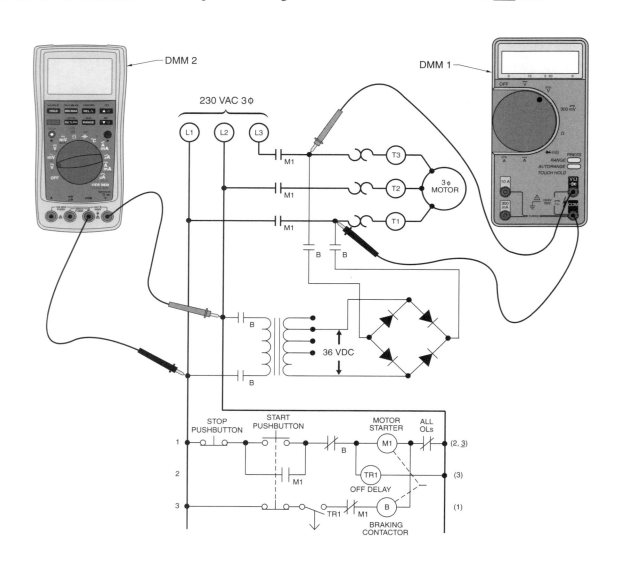

Preventive Maintenance and Troubleshooting

19 Applications

Application 19-1: Conveyor Drive Methods

Conveyor Drives

The two drive methods used to move products along a conveyor belt are the direct drive and roller drive methods. The direct drive method has the product riding directly on the driven belt and is used for light loads. The roller drive method has the belt driving rollers on which the product rides. The roller drive method is used for heavy loads. **See Conveyor Drive Methods.**

The right side of a conveyor is the right side when facing the forward direction of material travel. The two ends of a conveyor are identified by their relationship to the forward direction of material travel. The *tail end* of a conveyor is the end where material is fed. The *head end* of a conveyor is the end from which material is discharged. A drive unit is connected to the tail end pulley or head end pulley.

Belt Tension

To prevent slippage, a conveyor belt must have the correct tension. Proper belt tension is accomplished by adjusting the take-up pulley. A *take-up pulley* is a pulley that is used for correcting belt tension and is not connected to a drive.

Belt Tracking

Proper belt tracking depends on the alignment of the pulleys and rollers. A *pulley* is a revolving cylinder connected to a drive. A *roller* is a revolving cylinder not connected to a drive. As a conveyor belt moves, the belt should remain in the center of the pulleys and rollers. When a belt drifts to one side, the belt edge wears or folds up on the conveyor guard. Production downtime is required to replace or fix the belt.

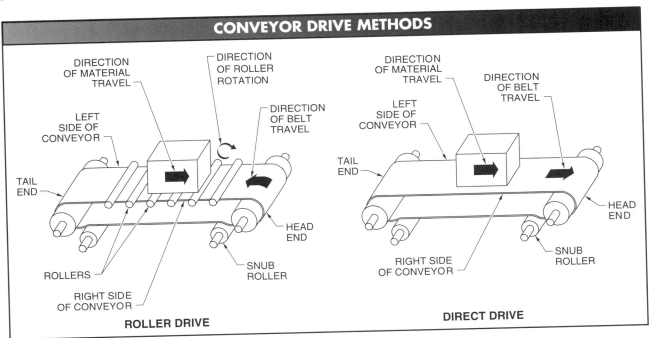

CONVEYOR DRIVE METHODS

271

A conveyor belt drifts toward the side where pulley and roller centers are not parallel and are the closest to each other. Conveyor belt tracking is adjusted by moving the pulley or snub roller. A *snub roller* is a roller not connected to a drive that is used to correct belt alignment. When aligning a belt, all adjustments should be slight, and time should be allowed for the belt to react to the adjustment.

Conveyor Adjustment — Head End Pulley

If a conveyor belt drifts to the right on the head end pulley during forward material travel, the right side of the head end snub roller is adjusted in the forward direction of material travel and/or the left side of the head end snub roller is adjusted in the reverse direction of material travel. **See Conveyors.**

Conveyor Adjustment — Tail End Pulley

If a conveyor belt drifts to the right on the tail end pulley during forward material travel, the right side of the tail end snub roller is adjusted in the reverse direction of material travel and/or the left side of the tail end snub roller is adjusted in the forward direction of material travel.

Conveyor Adjustment — Center Drive

If a conveyor belt drifts to the right side of the center drive and take-up pulleys, the right side of the snub roller is adjusted in the reverse direction of material travel and/or the left side of the snub roller is adjusted in the forward direction of material travel. **See Center Drive Conveyors.**

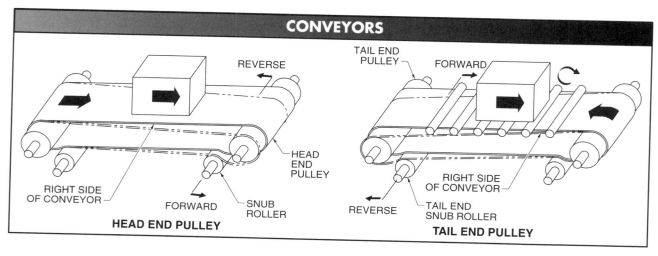

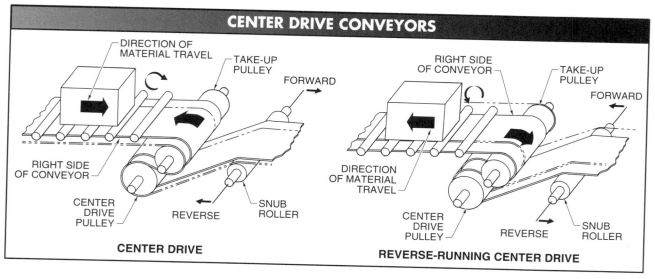

Application 19-2: Motor Coupling Selection

Conveyor Adjustment — Reverse-Running Center Drive

If a conveyor belt drifts to the right side of the center drive and take-up pulleys, the right side of the snub roller is adjusted in the forward direction of material travel and/or the left side of the snub roller is adjusted in the reverse direction of material travel.

Motor Couplings

A *motor coupling* is a device that connects a motor shaft to the equipment the motor is driving. A motor coupling allows a motor to operate the driven equipment, allows for a slight misalignment between the motor and the driven equipment, and allows for horizontal and axial movement of the shafts.

Misalignment

When connecting equipment, angular misalignment and parallel misalignment occur. *Angular misalignment* is misalignment when two shafts are not parallel. *Parallel misalignment* is misalignment when two shafts are parallel but not on the same line. **See Motor Couplings.**

Motor Coupling Ratings

Motor couplings are rated according to the amount of torque they can handle. Couplings are rated in inch-pounds (in-lb) or foot-pounds (ft-lb). The coupling torque rating must be correct for the application to prevent the coupling from bending or breaking. A bent coupling causes misalignment and vibration. A broken coupling prevents the motor from doing work.

Selecting Motor Couplings

The correct coupling is selected for an application by determining the nominal torque rating of the power source, determining the application service factor, calculating the coupling torque rating, selecting a coupling with an equal or greater torque rating, and

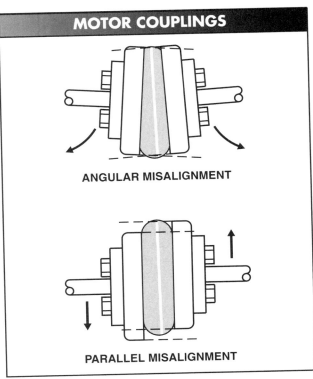

MOTOR COUPLINGS

ANGULAR MISALIGNMENT

PARALLEL MISALIGNMENT

ensuring that the coupling has the correct shaft size to fit the drive unit. To select the correct coupling for an application, apply the following procedure:

1. Determine the nominal torque rating of the power source (electric motor or other power source). The nominal torque rating is calculated, or found from a conversion table. To calculate the nominal torque rating of a motor in in-lb, apply the following formula:

$$T = \frac{HP \times 63,000}{rpm}$$

where
T = nominal torque rating (in in-lb)
HP = horsepower
$63,000$ = constant
rpm = speed (in revolutions per minute)

To calculate the nominal torque rating of a motor in ft-lb, apply the following formula:

$$T = \frac{HP \times 5252}{rpm}$$

where
T = nominal torque rating (in ft-lb)
HP = horsepower
5252 = constant
rpm = speed (in revolutions per minute)

To find the nominal torque rating of a motor using a conversion table, one end of a straightedge is placed on the rpm and the other end on the horsepower. The point where the straightedge crosses the torque scale is the torque rating. **See Horsepower to Torque Conversion.** For example, a 1 HP motor operates at 1750 rpm. To find the nominal torque rating in ft-lb, one end of a straightedge is placed on 1750 rpm and the other end at 1 HP. The straightedge crosses the torque scale at 3 ft-lb. The nominal torque rating is 3 ft-lb.

2. Determine application service factor. An *application service factor* is a multiplier that corrects for the operating conditions of a coupling. The greater the stress placed on a coupling, the larger the multiplier. By applying a multiplier, the size of the coupling is increased to adjust for severity of the load placed on the motor. **See Common Service Factors.**

3. Calculate the coupling torque rating by multiplying the nominal torque rating of the power source by the service factor of the application. Coupling torque rating is found by applying the following formula:

$$C_{TR} = N_{TR} \times SF$$

where
C_{TR} = coupling torque rating (in in-lb or ft-lb)
N_{TR} = nominal torque rating (in in-lb or ft-lb)
SF = service factor

4. Select a coupling with an equal or greater torque rating. Coupling torque ratings are given on coupling selection tables. **See Coupling Selections.**

5. Ensure the coupling has the correct shaft size to fit the drive unit. The exact size of a motor shaft can be determined by the motor frame number. Typical shaft sizes for motors from ¼ HP to 200 HP are ½″, ⅝″, ⅞″, 1⅛″, 1⅜″, 1⅝″, 1⅞″, 2⅛″, 2⅜″, 2⅞″, and 3⅜″.

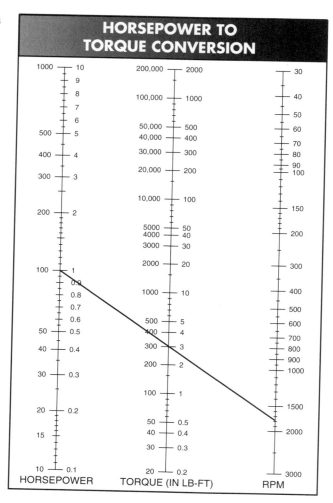

HORSEPOWER TO TORQUE CONVERSION

COMMON SERVICE FACTORS

Equipment	Service factor
Blowers	
Centrifugal	1.00
Vane	1.25
Compressors	
Centrifugal	1.25
Vane	1.50
Conveyors	
Uniformly loaded or fed	1.50
Heavy-duty	2.00
Elevators	
Bucket	2.00
Freight	2.25
Extruders	
Plastic	2.00
Metal	2.50
Fans	
Light-duty	1.00
Centrifugal	1.50

COUPLING SELECTIONS

Coupling number	Rated torque (in-lb)	Maximum shock torque (in-lb)
10-101-A	16	45
10-102-A	36	100
10-103-A	80	220
10-104-A	132	360
10-105-A	176	480
10-106-A	240	660
10-107-A	325	900
10-108-A	525	1450

Example: Motor Coupling Selection
A 1.5 HP motor operating at 1750 rpm is used in a heavy-duty conveyor. Select the correct coupling for the application.

1. Determine nominal torque rating.

$$T = \frac{HP \times 63,000}{rpm}$$

$$T = \frac{1.5 \times 63,000}{1750}$$

$$T = \frac{94,500}{1750}$$

$T = \mathbf{54\ in\text{-}lb}$

Note: 54 in-lb = 4.5 ft-lb (54 ÷ 12 = 4.5)

2. Determine application service factor.
The service factor for a heavy-duty conveyor is 2 (from Common Service Factors table).

3. Calculate the coupling torque rating.

$C_{TR} = N_{TR} \times SF$
$C_{TR} = 54 \times 2$
$C_{TR} = \mathbf{108\ in\text{-}lb}$

4. Select a coupling with an equal or greater torque rating.

The coupling with a torque rating equal to or greater than 108 in-lb is a 10-104-A coupling (from Coupling Selections table).

5. Ensure the coupling has the correct shaft size to fit the drive unit.

A 1.5 HP motor normally has a 145T frame. The shaft size of a 145T frame motor is ⅞". The coupling must have a bore size that would accept the ⅞" motor shaft.

Application 19-3: Extension Cord Selection

Extension Cords

Extension cords allow the use of appliances, tools, or other loads at a location away from a receptacle (outlet). Extension cords should only be used for delivering power on a temporary basis. Only extension cords that are in good working condition and of the proper type and size should be used.

When adding an extension cord to an electrical system, the extension cord conductor (wire) has a voltage drop across it because of the conductor's resistance. The resistance of the conductor is in series with the resistance of the load (lamp, power tool, etc.) connected to the extension cord and adds to the total resistance of the circuit. The total available voltage is divided between the conductor in the circuit and the load connected to the circuit. In an electrical system, the conductor should drop less than 3% of the applied voltage. The voltage applied to a load should be within +5% to −10% of the load's rating.

The amount of voltage dropped across an extension cord (or any length of conductor) is based on the conductor's size, length, and resistance, as well as the amount of current flowing through the conductor.

Extension Cord Selection Considerations

- The larger the diameter of a conductor, the less the total voltage drop across the conductor and the greater the voltage applied to the load.
- The lower the American Wire Gauge (AWG) number, the larger the diameter of the conductor. For example, an AWG #12 conductor is larger in diameter than an AWG #14 conductor.
- The shorter the total length of a conductor, the less the total voltage drop across the conductor and the greater the voltage applied to the load.
- The lower the resistance of a conductor (a copper conductor has less resistance than an aluminum conductor of the same AWG wire size), the less the total voltage drop across the conductor and the greater the voltage applied to the load.
- The lower the current draw by the load, the less the total voltage drop across the conductor and the greater the voltage applied to the load.
- The higher the voltage drop across a conductor, the higher the temperature of the conductor, since power (heat) is equal to the voltage drop times the amount of current flowing in the conductor.
- The higher the conductor temperature, the faster the conductor insulation breakdown.
- Loads that include electrical motors, such as power tools that are designed for operation on 115/120 VAC power supplies, draw high currents to deliver the required power. **See Power Tools.**

Extension cords are used with power tools because power tools are used in locations that are removed from a receptacle (outlet). Extension cord ratings include length, voltage (set by plug configuration), maximum current capacity, type (insulation), and intended usage (indoor/outdoor, indoor only, heavy-, light-, or medium-duty).

Manufacturers of power tools typically include a size and length chart that lists the maximum extension cord lengths for a given cord wire size and the expected amount of current draw. The charts ensure that by using the proper size extension cord, the tool is not damaged from low voltage and overheating. Using the proper size extension cord also provides a safe working environment. Size and length charts are used when extension cords are connected together to increase the length of the cord run. **See Selected Extension Cord Sizes and Lengths.**

SELECTED EXTENSION CORD SIZES AND LENGTHS*

Max. Amps @ 125 V	Extension Cord Length†				
	25	50	100	150	200
0 to 2	18	18	18	16	16
2 to 3	18	18	16	14	14
3 to 4	18	16	14	14	12
4 to 5	18	16	14	12	10
5 to 6	18	16	12	12	10
6 to 8	16	14	12	10	10
8 to 10	16	14	12	10	–
10 to 12	16	14	10	–	–
12 to 15	12	12	10	–	–
15 to 20	10	10	10	–	–

* AWG
† in ft

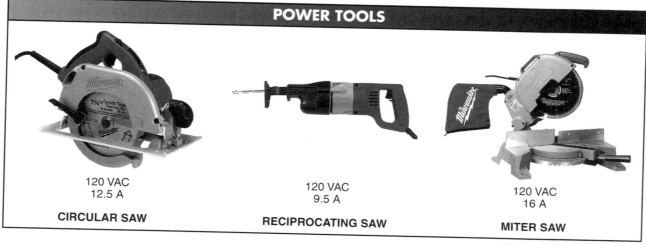

POWER TOOLS

CIRCULAR SAW — 120 VAC, 12.5 A

RECIPROCATING SAW — 120 VAC, 9.5 A

MITER SAW — 120 VAC, 16 A

Milwaukee Electric Tool Corp.

Application 19-4: Load Variations

Effect of Voltage Variation on Loads

All electrical loads (motors, lamps, computers, etc.) are rated to operate at a given voltage. The rated voltage may be a single voltage rating (12 VDC, 24 VAC, 115 VAC, etc.), a dual-voltage rating (115/230 VAC, etc.), or a voltage range (10 VDC – 30 VDC). Most electrical loads such as computers, lamps, hand-held electric tools, kitchen appliances, and office equipment have a single voltage rating. Large power-consuming electrical loads such as motors over ½ HP, electric ranges (ovens), refrigerators, and heaters over 1000 W have a dual-voltage rating. In all dual-voltage rated loads, the high voltage is preferred because the high voltage requires low current to produce the same power output. Electronic devices such as photoelectric switches, proximity switches, and solid-state timers often have a voltage range. **See Supply Voltage Ratings.**

Most electrical loads operate satisfactorily with a voltage variation +5% to –10% of the load's rated operating voltage. However, applied voltages that are outside of the load's operating range will affect any load. **See Voltage Variations.**

Although voltage variation charts show what occurs to operating characteristics such as torque, current, temperature, lumens (light output), etc., for different load types, applied voltages should still fall within a limited range. The old acceptable voltage variation range was ±10% of the load's rated voltage. The new acceptable voltage variation range of +5% to –10% applies to most new loads such as computers. The reason for this is that, in general, a high voltage causes more damage than a low voltage. When a low voltage is applied to a load, lamps dim, heating elements produce less heat, and computers may lose memory or fail to operate, but when the voltage is brought back to normal, the device operates properly. However, when a high voltage is applied to a load, the device can be permanently damaged. Returning the voltage to normal does not repair the damage.

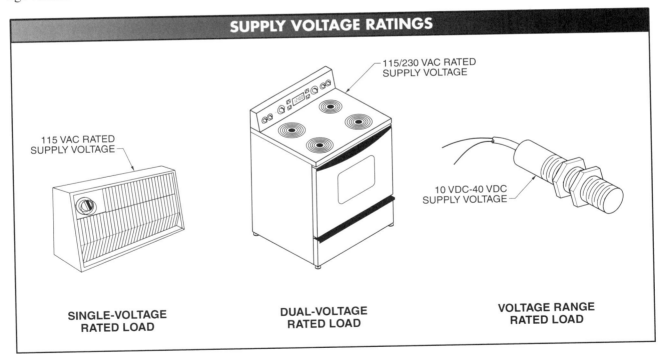

VOLTAGE VARIATIONS

VOLTAGE VARIATIONS – AC MOTORS

Performance Characteristics	10% below Rated Voltage	10% above Rated Voltage
Starting current	−10% to −12%	+10% to +12%
Full-load current	+11%	−7%
Motor torque	−20% to −25%	+20% to +25%
Motor efficiency	Little change	Little change
Speed	−1.5%	+1%
Temperature rise	+6°C to +7°C	−3°C to −4°C

VOLTAGE VARIATIONS – DC MOTORS

Performance Characteristics	10% below Rated Voltage		10% above Rated Voltage	
	Shunt	Compound	Shunt	Compound
Starting torque	−15%	−15%	+15%	+15%
Speed	−5%	−6%	+5%	+6%
Current	+12%	+12%	−8%	−8%
Field temperature	Decreases	Decreases	Increases	Increases
Armature temperature	Increases	Increases	Decreases	Decreases
Commutator temperature	Increases	Increases	Decreases	Decreases

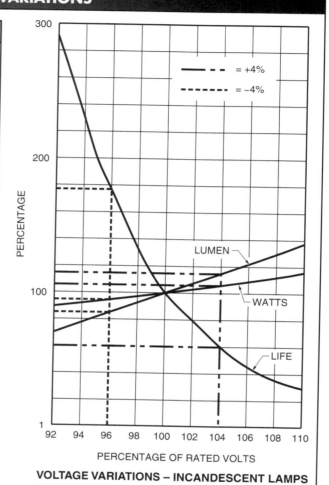

VOLTAGE VARIATIONS – INCANDESCENT LAMPS

Preventive Maintenance and Troubleshooting

19 Activities

Activity 19-1: Conveyor Drive Methods

Identify the roller that must be adjusted for proper belt tracking. Show the direction of adjustment on the drawing.

_____ 1. Roller to be adjusted.   _____ 2. Roller to be adjusted.

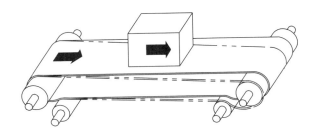
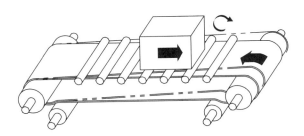

_____ 3. Roller to be adjusted.   _____ 4. Roller to be adjusted.

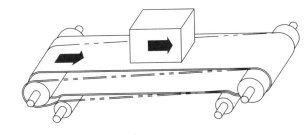
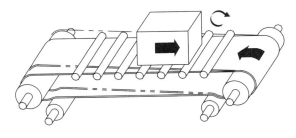

_____ 5. Roller to be adjusted.   _____ 6. Roller to be adjusted.

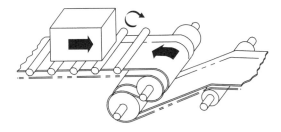
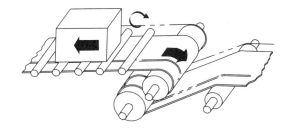

Activity 19-2: Motor Coupling Selection

Calculate the coupling torque rating and select a coupling using Coupling Selections on page 275.

_____ 1. A heavy-duty conveyor has a 2 HP motor that turns at 1800 rpm. The coupling torque rating is ___ in-lb.

_____ 2. A uniformly loaded standard conveyor has a 5 HP motor that turns at 1200 rpm. The coupling torque rating is ___ in-lb.

_____ 3. A printing press has an 8.25 HP motor that turns at 1800 rpm. The coupling torque rating is ___ in-lb.

_____ 4. A textile loom has a 2.5 HP motor that turns at 900 rpm. The coupling torque rating is ___ ft-lb.

_____ 5. A punch press has a 3.75 HP motor that turns at 1200 rpm. The coupling torque rating is ___ ft-lb.

_____ 6. Coupling number ___ is used for an application with a 205 in-lb torque rating.

_____ 7. Coupling number ___ is used for an application with a 30 in-lb torque rating.

_____ 8. Coupling number ___ is used for an application with a 475 in-lb torque rating.

_____ 9. Coupling number ___ is used for an application with a 154 in-lb torque rating.

_____ 10. Coupling number ___ is used for an application with a 17 in-lb torque rating.

Activity 19-3: Extension Cord Selection

Determine the minimum acceptable extension cord that can be used for each application from Extension Cords on page 281 and from Power Tools on page 276. Note: If the extension cord length exceeds the listings on the manufacturer listed extension cord size and length chart for a given load current rating, list the answer as none in place of a extension cord model number.

_____ 1. Model number ___ is the minimum acceptable extension cord that can be used to deliver power to a circular saw 25′ from a receptacle.

_____ 2. Model number ___ is the minimum acceptable extension cord that can be used to deliver power to a circular saw 50′ from a receptacle.

_____ 3. Model number ___ is the minimum acceptable extension cord that can be used to deliver power to a circular saw 100′ from a receptacle.

_____   4. Model number ___ is the minimum acceptable extension cord that can be used to deliver power to a circular saw 200′ from a receptacle.

_____   5. Model number ___ is the minimum acceptable extension cord that can be used to deliver power to a reciprocating saw 25′ from a receptacle.

_____   6. Model number ___ is the minimum acceptable extension cord that can be used to deliver power to a reciprocating saw 50′ from a receptacle.

_____   7. Model number ___ is the minimum acceptable extension cord that can be used to deliver power to a reciprocating saw 100′ from a receptacle.

_____   8. Model number ___ is the minimum acceptable extension cord that can be used to deliver power to a reciprocating saw 200′ from a receptacle.

_____   9. Model number ___ is the minimum acceptable extension cord that can be used to deliver power to a miter saw 25′ from a receptacle.

_____ 10. Model number ___ is the minimum acceptable extension cord that can be used to deliver power to a miter saw 100′ from a receptacle.

| EXTENSION CORDS ||
Model Number	Extension Cord Type
18-25	25′ AWG #18, 120 VAC
18-50	50′ AWG #18, 120 VAC
18-100	100′ AWG #18, 120 VAC
16-25	25′ AWG #16, 120 VAC
16-50	50′ AWG #16, 120 VAC
16-100	100′ AWG #16, 120 VAC
14-25	25′ AWG #14, 120 VAC
14-50	50′ AWG #14, 120 VAC
14-100	100′ AWG #14, 120 VAC
14-150	150′ AWG #14, 120 VAC
12-25	25′ AWG #12, 120 VAC
12-50	50′ AWG #12, 120 VAC
12-100	100′ AWG #12, 120 VAC
12-150	150′ AWG #12, 120 VAC
10-25	25′ AWG #10, 120 VAC
10-50	50′ AWG #10, 120 VAC
10-100	100′ AWG #10, 120 VAC

Activity 19-4: Load Variations

Determine each value using Voltage Variations – AC Motors on page 278.

_____ 1. The starting current of the motor is ___% based on the reading of DMM 1 and the motor rating.

_____ 2. The starting current of the motor is ___% based on the reading of DMM 2 and the motor rating.

_____ 3. The torque of the motor is ___% based on the reading of DMM 1 and the motor rating.

_____ 4. The torque of the motor is ___% based on the reading of DMM 2 and the motor rating.

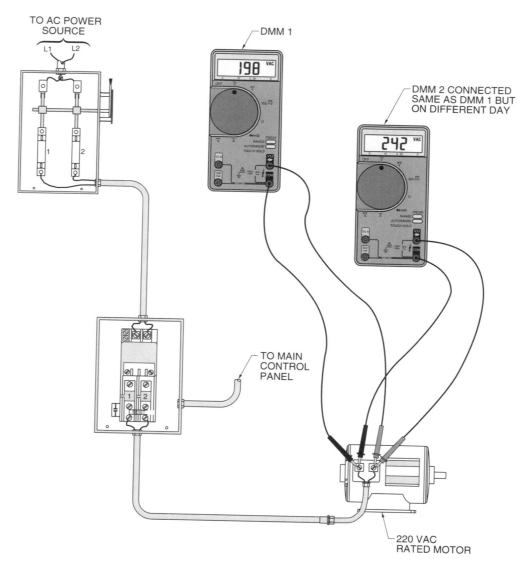

Determine each value using Effects of Voltage Variations – DC Motors on page 278.

_____ 5. The current of the motor is ___% based on the reading of DMM 1 and the motor rating.

_____ 6. The current of the motor is ___% based on the reading of DMM 2 and the motor rating.

_____ 7. The speed of the motor is ___% based on the reading of DMM 1 and the motor rating.

_____ 8. The speed of the motor is ___% based on the reading of DMM 2 and the motor rating.

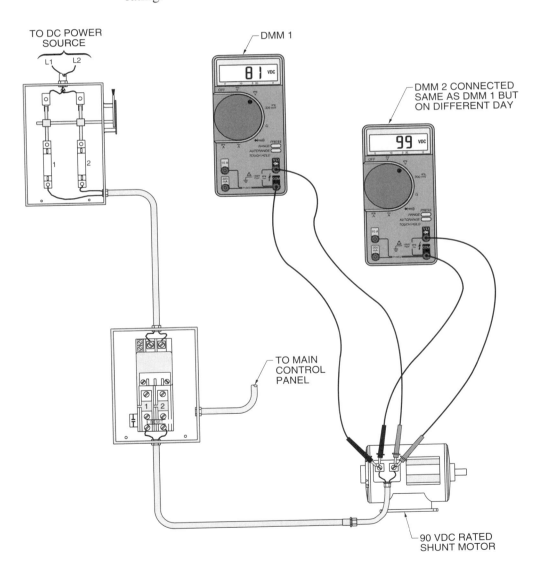

Determine each value using Voltage Variations – Incandescent Lamps on page 278.

_____ 9. The lamp life is ___ % of the rated lamp life based on the reading of DMM 1 and the lamp rating.

_____ 10. The lamp life is ___ % of the rated lamp life based on the reading of DMM 2 and the lamp rating.

_____ 11. The lumen output is ___ % of the rated lumen output based on the reading of DMM 1 and the lamp rating.

_____ 12. The lumen output is ___ % of the rated lumen output based on the reading of DMM 2 and the lamp rating.

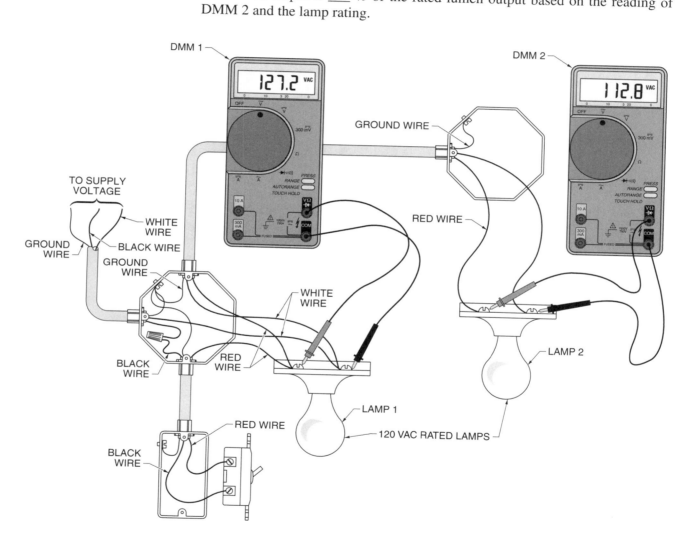

Appendix

Average Wattage Requirement Guide	286
Heater Trip Characteristics	287
Heater Selections	288
Full-load Currents — 3φ, AC Induction Motors	289
Full-load Currents — 1φ, AC Motors	289
Full-load Currents — DC Motors	289
Control Ratings	290
1φ Motors and Circuits	292
3φ, 230 V Motors and Circuits — 240 V System	293
3φ, 460 V Motors and Circuits — 480 V System	294
DC Motors and Circuits	295
Hazardous Locations	296
Electronic Symbols	297
Industrial Electrical Symbols	298
Electrical/Electronic Abbreviations/Acronyms	301
Nonlocking Wiring Devices	302
Horsepower to Torque Conversion	303
Common Service Factors	304
Certificates of Completion	305

AVERAGE WATTAGE REQUIREMENT GUIDE

Electrical Load	Operating Power*	Starting Power*	Electrical Load	Operating Power*	Starting Power*
Construction site			**Household**		
Air Compressor (½ HP)	1500	5500	Electric Clothes Dryer	6000	7500
Air Compressor (1 HP)	3000	11,000	Refrigerator (standard size)	800	2000
Electric Welder (200 A)	9000	9000	Dishwasher	1100	2100
High-Pressure Washer (1 HP)	1200	3600	Radio	50-200	50-200
Circular Saw (7¼″)	1400	2300	Sump Pump (⅓ HP)	800	1300
Table Saw (10″)	1800	4500	Sump Pump (½ HP)	1050	2150
Hand Drill (½″)	600	800	Furnace (gas/oil, ¼ HP electric fan)	600	1000
Grinder (4½″)	750	950	Furnace (gas/oil, ½ HP electric fan)	900	2400
Grinder (6″)	1000	1300	Air Conditioning Unit	800-1500	1200-2000
Grinder (9″)	2300	3000	Home Security System	100	100
Hand Jigsaw	650	850	Iron	1200	1200
Reciprocating Saw (7″ blade)	1150	1600	Space Heater	1500	1500
Sander (⅓ sandpaper sheet size)	350	550	Heating Pad	50-75	50-75
Sander (½ sandpaper sheet size)	450	650	Hot Water Heater	2500-5000	2500-5000
Battery Charger (15 A, no boost)	375	375			
Battery Charger (60 A, no boost)	1500	1500	**Office**		
Battery Charger (100 A, no boost)	2500	2500	Computer (laptop)	250	300
			Computer (desktop)	600	800
Household			Monitor	250	250
Lights (as indicated on bulb)			Printer	400	600
100 W bulb	100	100	Fax machine	500	700
60 W bulb	60	60			
Toaster	1100	1100	**Motors (single-phase)**		
Can Opener	150	250	⅛ HP	190	400
Clock	15	15	¼ HP	375	825
Television (19″ color)	300	300	⅓ HP	500	1200
Microwave Oven (standard size)	700	900	½ HP	750	1650
Coffeemaker	1700	1700	¾ HP	1125	2475
Electric Skillet (hot plate)	1500	1500	1 HP	1500	3300
Electric Range (8″ element)	2100	2100	1½ HP	2250	4950
Electric Oven	6000	6000			

* in W

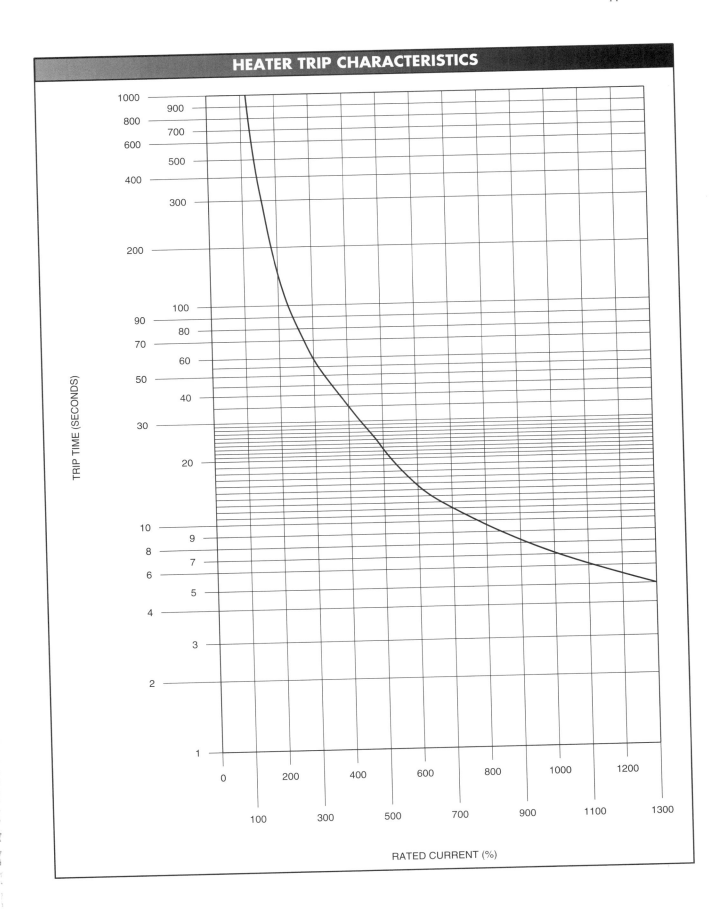

HEATER SELECTIONS

Heater Number	Full Load Current (A)				
	Size 0	Size 1	Size 2	Size 3	Size 4
10	0.20	0.20	—	—	—
11	0.22	0.22	—	—	—
12	0.24	0.24	—	—	—
13	0.27	0.27	—	—	—
14	0.30	0.30	—	—	—
15	0.34	0.34	—	—	—
16	0.37	0.37	—	—	—
17	0.41	0.41	—	—	—
18	0.45	0.45	—	—	—
19	0.49	0.49	—	—	—
20	0.54	0.54	—	—	—
21	0.59	0.59	—	—	—
22	0.65	0.65	—	—	—
23	0.71	0.71	—	—	—
24	0.78	0.78	—	—	—
25	0.85	0.85	—	—	—
26	0.93	0.93	—	—	—
27	1.02	1.02	—	—	—
28	1.12	1.12	—	—	—
29	1.22	1.22	—	—	—
30	1.34	1.34	—	—	—
31	1.48	1.48	—	—	—
32	1.62	1.62	—	—	—
33	1.78	1.78	—	—	—
34	1.96	1.96	—	—	—
35	2.15	2.15	—	—	—
36	2.37	2.37	—	—	—
37	2.60	2.60	—	—	—
38	2.86	2.86	—	—	—
39	3.14	3.14	—	—	—
40	3.45	3.45	—	—	—
41	3.79	3.79	—	—	—
42	4.17	4.17	—	—	—
43	4.58	4.58	—	—	—
44	5.03	5.03	—	—	—
45	5.53	5.53	—	—	—
46	6.08	6.08	—	—	—

HEATER SELECTIONS

Heater Number	Full Load Current (A)				
	Size 0	Size 1	Size 2	Size 3	Size 4
47	6.68	6.68	—	—	—
48	7.21	7.21	—	—	—
49	7.81	7.81	7.89	—	—
50	8.46	8.46	8.57	—	—
51	9.35	9.35	9.32	—	—
52	10.00	10.00	10.1	—	—
53	10.7	10.7	11.0	12.2	—
54	11.7	11.7	12.0	13.3	—
55	12.6	12.6	12.9	14.3	—
56	13.9	13.9	14.1	15.6	—
57	15.1	15.1	15.5	17.2	—
58	16.5	16.5	16.9	18.7	—
59	18.0	18.0	18.5	20.5	—
60	—	19.2	20.3	22.5	23.8
61	—	20.4	21.8	24.3	25.7
62	—	21.7	23.5	26.2	27.8
63	—	23.1	25.3	28.3	30.0
64	—	24.6	27.2	30.5	32.5
65	—	26.2	29.3	33.0	35.0
66	—	27.8	31.5	36.0	38.0
67	—	—	33.5	39.0	41.0
68	—	—	36.0	42.0	44.5
69	—	—	38.5	45.5	48.5
70	—	—	41.0	49.5	52
71	—	—	43.0	53	57
72	—	—	46.0	58	61
73	—	—	—	63	67
74	—	—	—	68	72
75	—	—	—	73	77
76	—	—	—	78	84
77	—	—	—	83	91
78	—	—	—	88	97
79	—	—	—	—	103
80	—	—	—	—	111
81	—	—	—	—	119
82	—	—	—	—	127
83	—	—	—	—	133

FULL-LOAD CURRENTS — 3φ, AC INDUCTION MOTORS

Motor rating (HP)	Current (A)			
	208 V	230 V	460 V	575 V
1/4	1.11	.96	.48	.38
1/3	1.34	1.18	.59	.47
1/2	2.2	2.0	1.0	.8
3/4	3.1	2.8	1.4	1.1
1	4.0	3.6	1.8	1.4
1 1/2	5.7	5.2	2.6	2.1
2	7.5	6.8	3.4	2.7
3	10.6	9.6	4.8	3.9
5	16.7	15.2	7.6	6.1
7 1/2	24.0	22.0	11.0	9.0
10	31.0	28.0	14.0	11.0
15	46.0	42.0	21.0	17.0
20	59	54	27	22
25	75	68	34	27
30	88	80	40	32
40	114	104	52	41
50	143	130	65	52
60	169	154	77	62
75	211	192	96	77
100	273	248	124	99
125	343	312	156	125
150	396	360	180	144
200	—	480	240	192
250	—	602	301	242
300	—	—	362	288
350	—	—	413	337
400	—	—	477	382
500	—	—	590	472

FULL-LOAD CURRENTS — 1φ, AC MOTORS

Motor rating (HP)	Current (A)	
	115 V	230 V
1/6	4.4	2.2
1/4	5.8	2.9
1/3	7.2	3.6
1/2	9.8	4.9
3/4	13.8	6.9
1	16	8
1 1/2	20	10
2	24	12
3	34	17
5	56	28
7 1/2	80	40

FULL-LOAD CURRENTS — DC MOTORS

Motor rating (HP)	Current (A)	
	120 V	240 V
1/4	3.1	1.6
1/3	4.1	2.0
1/2	5.4	2.7
3/4	7.6	3.8
1	9.5	4.7
1 1/2	13.2	6.6
2	17	8.5
3	25	12.2
5	40	20
7 1/2	48	29
10	76	38

CONTROL RATINGS

Size	Load (V)	Maximum HP Normal duty 1φ	Maximum HP Normal duty 3φ	Maximum HP Plugging & jogging duty 1φ	Maximum HP Plugging & jogging duty 3φ	Cont amps	Service limit amps	Tungsten & ballast type lamp amps 480 V max	Resistance heating (kW) 1φ	Resistance heating (kW) 3φ	Transformer switching 50–60 Hz kVA rating inrush peak time Continuous amps 20 times 1φ	20 times 3φ	20–40 times 1φ	20–40 times 3φ	Capacitor kVA switching rating 3φ kVAR
00	115	½	—	—	—	9	11	—	1.15	2.0	—	—	—	—	—
00	200	—	1½	—	—	9	11	—	2.0	3.46	—	—	—	—	—
00	230	1	1½	—	—	9	11	—	2.3	4.0	—	—	—	—	—
00	380	—	1½	—	—	9	11	—	—	6.5	—	—	—	—	—
00	460	—	2	—	—	9	11	—	4.6	8.0	—	—	—	—	—
00	575	—	2	—	—	9	11	—	5.8	10.0	—	—	—	—	—
0	115	1	—	½	—	18	21	20	2.3	4.0	0.6	—	0.3	—	—
0	200	—	3	—	1½	18	21	20	4.0	6.92	—	1.8	—	0.9	—
0	230	2	3	1	1½	18	21	20	4.6	8.0	1.2	2.1	0.6	1.0	—
0	380	—	5	—	1½	18	21	20	—	13.1	—	—	—	—	—
0	460	—	5	—	2	18	21	20	9.2	15.9	2.4	4.2	1.2	2.1	—
0	575	—	5	—	2	18	21	—	11.5	19.9	3.0	5.2	1.5	2.6	—
1	115	2	—	1	—	27	32	30	3.5	6.0	1.2	—	0.6	—	—
1	200	—	7½	—	3	27	32	30	6	10.4	—	3.6	—	1.8	—
1	230	3	7½	2	3	27	32	30	6.9	11.9	2.4	4.3	1.2	2.1	—
1	380	—	10	—	5	27	32	30	—	19.7	—	—	—	—	—
1	460	—	10	—	5	27	32	30	13.8	23.9	4.9	8.5	2.5	4.3	—
1	575	—	10	—	5	27	32	—	17.3	29.8	6.2	11.0	3.1	5.3	—
1P	115	3	—	1½	—	35	42	45	5.8	—	—	—	—	—	—
1P	230	5	—	3	—	35	42	45	11.5	—	—	—	—	—	—
1¾	115	—	—	—	—	40	40	45	5.8	9.9	1.6	—	0.8	—	—
1¾	200	—	10	—	5	40	40	45	10	17.3	—	4.9	—	2.4	—
1¾	230	—	10	—	5	40	40	45	11.5	19.9	3.2	5.75	1.6	2.8	—
1¾	380	—	15	—	7½	40	40	45	—	32.9	—	—	—	—	—
1¾	460	—	15	—	7½	40	40	45	23	39.8	6.6	11.2	3.3	5.7	—
1¾	575	—	15	—	7½	40	40	—	28.8	49.7	8.1	14.5	4.1	7.1	—
2	115	3	—	2	—	45	52	60	8.1	13.9	2.1	—	1.0	—	—
2	200	—	10	—	7½	45	52	60	14	24.2	—	6.3	—	3.1	—
2	230	7½	15	5	1	45	52	60	16.1	27.8	4.1	7.2	2.1	3.6	8
2	380	—	25	—	1	45	52	60	—	46.0	—	—	—	—	—
2	460	—	25	—	15	45	52	60	32.2	55.7	8.3	14	4.2	7.2	16
2	575	—	25	—	15	45	52	—	40.3	69.6	10.0	18	5.2	8.9	20
2½	115	5	—	—	—	60	65	75	10.4	17.9	3.1	—	1.5	—	—
2½	200	—	10	—	10	60	65	75	18	31.1	—	9.1	—	4.6	—
2½	230	10	20	—	10	60	65	75	20.7	35.8	6.1	10.6	3.1	5.3	17.5
2½	380	—	30	—	20	60	65	75	—	59.2	—	—	—	—	—
2½	460	—	30	—	20	60	65	75	41.4	71.6	12	21	6.1	10.6	34.5
2½	575	—	30	—	20	60	65	—	51.8	89.5	15	26.5	7.6	13.4	43.5

continued

continued

CONTROL RATINGS

Size	Load (V)	Maximum HP Normal duty 1φ	Maximum HP Normal duty 3φ	Maximum HP Plugging & jogging duty 1φ	Maximum HP Plugging & jogging duty 3φ	Cont amps	Service limit amps	Tungsten & ballast type lamp amps 480 V max	Resistance heating (kW) 1φ	Resistance heating (kW) 3φ	Transformer switching 50–60 Hz kVA rating inrush peak time Continuous amps 20 times 1φ	20 times 3φ	20–40 times 1φ	20–40 times 3φ	Capacitor kVA switching rating 3φ kVAR
3	115	7½	—	—	—	90	104	100	14.4	24.8	4.1	—	2.0	—	—
3	200	—	25	—	15	90	104	100	25	43.3	—	12	—	6.1	—
3	230	15	30	—	20	90	104	100	28.8	50.0	8.1	14	4.1	7.0	27
3	380	—	50	—	30	90	104	100	—	82.2	—	—	—	—	—
3	460	—	50	—	30	90	104	100	57.5	99.4	16	28	8.1	14	53
3	575	—	50	—	30	90	104	—	71.9	124	20	35	10	18	67
3½	115	—	—	—	—	115	125	150	18.4	31.8	—	—	—	—	—
3½	200	—	30	—	20	115	125	150	32	55.4	—	16	—	8	—
3½	230	—	60	—	25	115	125	150	36.8	63.7	11	18.5	5.4	9.5	33.5
3½	380	—	60	—	30	115	125	150	—	105	—	—	—	—	—
3½	460	—	75	—	40	115	125	150	73.6	127	21.5	37.5	11.0	18.5	66.5
3½	575	—	75	—	40	115	125	—	92	159	37	47	13.5	23.5	83.5
4	200	—	40	—	25	135	156	200	39	67.5	—	20	—	10	—
4	230	—	50	—	30	135	156	200	44.9	77.6	14	23	6.8	12	40
4	380	—	75	—	50	135	156	200	—	128	—	—	—	—	—
4	460	—	100	—	60	135	156	200	89.7	155	27	47	14	23	80
4	575	—	100	—	60	135	156	—	112	194	34	59	17	29	100
4½	200	—	50	—	30	210	225	250	53	91.7	—	30.5	—	15	—
4½	230	—	75	—	40	210	225	250	60.9	105	20.5	35	10.4	18	60
4½	380	—	100	—	75	210	225	250	—	174	—	—	—	—	—
4½	460	—	150	—	100	210	225	250	122	211	40.5	70.5	20.5	35	120
4½	575	—	150	—	100	210	225	—	152	264	51	88	25.5	44	150

1φ MOTORS AND CIRCUITS

1		2		3	4	5				6	
Size of motor		Motor overload protection — Low-peak or Fusetron®		Switch 115% minimum or HP rated or fuse holder size	Minimum size of starter	Controller termination temperature rating				Minimum size of copper wire and trade conduit	
						60°C		75°C			
HP	Amp	Motor less than 40°C or greater than 1.15 SF (Max fuse 125%)	All other motors (Max fuse 115%)			TW	THW	TW	THW	Wire size (AWG or kcmil)	Conduit (inches)
115 V (120 V system)											
1/6	4.4	5	5	30	00	•	•	•	•	14	1/2
1/4	5.8	7	6 1/4	30	00	•	•	•	•	14	1/2
1/3	7.2	9	8	30	00	•	•	•	•	14	1/2
1/2	9.8	12	10	30	00	•	•	•	•	14	1/2
3/4	13.8	15	15	30	00	•	•	•	•	14	1/2
1	16	20	17 1/2	30	00	•	•	•	•	14	1/2
1 1/2	20	25	20	30	01	•	•	•	•	12	1/2
2	24	30	25	30	01	•	•	•	•	10	1/2
230 V (240 V system)											
1/6	2.2	2 1/2	2 1/2	30	00	•	•	•	•	14	1/2
1/4	2.9	3 1/2	3 2/10	30	00	•	•	•	•	14	1/2
1/3	3.6	4 1/2	4	30	00	•	•	•	•	14	1/2
1/2	4.9	5 6/10	5 6/10	30	00	•	•	•	•	14	1/2
3/4	6.9	8	7 1/2	30	00	•	•	•	•	14	1/2
1	8	10	9	30	00	•	•	•	•	14	1/2
1 1/2	10	12	10	30	0	•	•	•	•	14	1/2
2	12	15	12	30	0	•	•	•	•	14	1/2
3	17	20	17 1/2	30	1	•	•	•	•	12	1/2
5	28	35	30	60	2		•			8	3/4
						•		•		8	1/2
									•	10	1/2
7 1/2	40	50	45	60	2	•	•	•		6	3/4
									•	8	3/4
										8	1/2
10	50	60	50	60	3	•	•	•		4	1
										4	3/4
									•	6	3/4

3φ, 230 V MOTORS AND CIRCUITS — 240 V SYSTEM

1		2		3	4	5				6	
Size of motor		Motor overload protection Low-peak or Fusetron®				Controller termination temperature rating				Minimum size of copper wire and trade conduit	
						60°C		75°C			
HP	Amp	Motor less than 40°C or greater than 1.15 SF (Max fuse 125%)	All other motors (Max fuse 115%)	Switch 115% minimum or HP rated or fuse holder size	Minimum size of starter	TW	THW	TW	THW	Wire size (AWG or kcmil)	Conduit (inches)
½	2	2½	2¼	30	00	•	•	•	•	14	½
¾	2.8	3½	3 2/10	30	00	•	•	•	•	14	½
1	3.6	4½	4	30	00	•	•	•	•	14	½
1½	5.2	6¼	5 6/10	30	00	•	•	•	•	14	½
2	6.8	8	7½	30	0	•	•	•	•	14	½
3	9.6	12	10	30	0	•	•	•	•	14	½
5	15.2	17½	17½	30	1	•	•	•	•	14	½
7½	22	25	25	30	1	•	•	•	•	10	½
10	28	35	30*	60	2	•	•	•		8	¾
									•	10	½
15	42	50	45	60	2	•	•	•	•	6	1
										6	¾
20	54	60*	60*	100	3	•	•	•	•	4	1
25	68	80	75	100	3	•	•			3	1¼
								•		3	1
									•	4	1
30	80	100	90	100	3	•	•	•		1	1¼
									•	3	1¼
40	104	125	110	200	4	•	•	•		2/0	1½
									•	1	1¼
50	130	150	150	200	4	•	•	•		3/0	2
									•	2/0	1½
75	192	225	200*	400	5	•	•	•		300	2½
									•	250	2½
100	248	300	250	400	5	•	•	•		500	3
									•	350	2½
150	360	450	400*	600	6	•	•	•		300-2/φ*	2-2½*
									•	4/0-2/φ*	2-2*

* fuse reducers required

3φ, 460 V MOTORS AND CIRCUITS — 480 V SYSTEM

1		2		3	4	5				6	
Size of motor		Motor overload protection Low-peak or Fusetron®				Controller termination temperature rating				Minimum size of copper wire and trade conduit	
						60°C		75°C			
HP	Amp	Motor less than 40°C or greater than 1.15 SF (Max fuse 125%)	All other motors (Max fuse 115%)	Switch 115% minimum or HP rated or fuse holder size	Minimum size of starter	TW	THW	TW	THW	Wire size (AWG or kcmil)	Conduit (inches)
½	1	1¼	1⅛	30	00	•	•	•	•	14	½
¾	1.4	1⁶⁄₁₀	1⁶⁄₁₀	30	00	•	•	•	•	14	½
1	1.8	2¼	2	30	00	•	•	•	•	14	½
1½	2.6	3²⁄₁₀	2⁶⁄₁₀	30	00	•	•	•	•	14	½
2	3.4	4	3½	30	00	•	•	•	•	14	½
3	4.8	5⁶⁄₁₀	5	30	0	•	•	•	•	14	½
5	7.6	9	8	30	0	•	•	•	•	14	½
7½	11	12	12	30	1	•	•	•	•	14	½
10	14	17½	15	30	1	•	•	•	•	14	½
15	21	25	20	30	2	•	•	•	•	10	½
20	27	30	30	60	2	•	•	•		8	¾
									•	10	½
25	34	40	35	60	2	•	•	•		6	1
									•	8	¾
30	40	50	45	60	3	•	•	•		6	1
									•	8	¾
40	52	60	60	100	3	•	•	•		4	1
									•	6	1
50	65	80	70	100	3	•	•	•		3	1¼
									•	4	1
60	77	90	80	100	4	•	•	•		1	1¼
									•	3	1¼
75	96	110	110	200	4	•	•	•		1/0	1½
									•	1	1¼
100	124	150	125	200	4	•	•	•		3/0	2
									•	2/0	1½
125	156	175	175	200	5	•	•	•		4/0	2
									•	3/0	2
150	180	225	200	400	5	•	•	•		300	2½
									•	4/0	2
200	240	300	250	400	5	•	•	•		500	3
									•	350	2½
250	302	350	325	400	6	•	•	•		4/0-2/φ*	2-2*
									•	3/0-2/φ*	2-2*
300	361	450	400	600	6	•	•	•		300-2/φ*	2-1½*
									•	4/0-2/φ*	2-2*

* two sets of multiple conductors and two runs of conduit required

DC MOTORS AND CIRCUITS

1	2		3	4	5				6	
Size of motor	Motor overload protection				Controller termination temperature rating				Minimum size of copper wire and trade conduit	
	Low-peak or Fusetron®				60°C		75°C			
	Motor less than 40°C or greater than 1.15 SF (Max fuse 125%)	All other motors (Max fuse 115%)	Switch 115% minimum or HP rated or fuse holder size	Minimum size of starter	TW	THW	TW	THW	Wire size (AWG or kcmil)	Conduit (inches)
HP / Amp										
90 V										
¼ / 4.0	5	4½	30	0	•	•	•	•	14	½
⅓ / 5.2	6¼	5⁶⁄₁₀	30	0	•	•	•	•	14	½
½ / 6.8	8	7½	30	0	•	•	•	•	14	½
¾ / 9.6	12	10	30	0	•	•	•	•	14	½
1 / 12.2	15	12	30	0	•	•	•	•	14	½
120 V										
¼ / 3.1	3½	3½	30	0	•	•	•	•	14	½
⅓ / 4.1	5	4½	30	0	•	•	•	•	14	½
½ / 5.4	6¼	6	30	0	•	•	•	•	14	½
¾ / 7.6	9	8	30	0	•	•	•	•	14	½
1 / 9.5	10	10	30	0	•	•	•	•	14	½
1½ / 13.2	15	15	30	1	•	•	•	•	14	½
2 / 17	20	17½	30	1	•	•	•	•	12	½
5 / 40	50	45	60	2	•	•	•		6	¾
								•	8	¾
									8	½
10 / 76	90	80	100	3	•	•	•		2	1
								•	3	1
180 V										
¼ / 2	2½	2¼	30	0	•	•	•	•	14	½
⅓ / 2.6	3²⁄₁₀	2⁸⁄₁₀	30	0	•	•	•	•	14	½
½ / 3.4	4	3½	30	0	•	•	•	•	14	½
¾ / 4.8	6	5	30	0	•	•	•	•	14	½
1 / 6.1	7½	7	30	0	•	•	•	•	14	½
1½ / 8.3	10	9	30	1	•	•	•	•	14	½
2 / 10.8	12	12	30	1	•	•	•	•	14	½
3 / 16	20	17½	30	1	•	•	•	•	12	½
10 / 27	30*	30*	60	1		•	•		8	½
						•			8	¾
								•	10	½

* two sets of multiple conductors and two runs of conduit required

HAZARDOUS LOCATIONS

Hazardous Location – A location where there is an increased risk of fire or explosion due to the presence of flammable gases, vapors, liquids, combustible dusts, or easily-ignitable fibers or flyings.

Location – A position or site.

Flammable – Capable of being easily ignited and of burning quickly.

Gas – A fluid (such as air) that has no independent shape or volume but tends to expand indefinitely.

Vapor – A substance in the gaseous state as distinguished from the solid or liquid state.

Liquid – A fluid (such as water) that has no independent shape but has a definite volume. A liquid does not expand indefinitely and is only slightly compressible.

Combustible – Capable of burning.

Ignitable – Capable of being set on fire.

Fiber – A thread or piece of material.

Flyings – Small particles of material.

Dust – Fine particles of matter.

Classes	Likelihood that a flammable or combustible concentration is present
I	Sufficient quantities of flammable gases and vapors present in air to cause an explosion or ignite hazardous materials
II	Sufficient quantities of combustible dust are present in air to cause an explosion or ignite hazardous materials
III	Easily ignitable fibers or flyings are present in air, but not in a sufficient quantity to cause an explosion of ignite hazardous materials

Divisions	Location containing hazardous substances
1	Hazardous location in which hazardous substance is normally present in air in sufficient quantities to cause an explosion or ignite hazardous materials
2	Hazardous location in which hazardous substance is not normally present in air in sufficient quantities to cause an explosion or ignite hazardous materials

Groups	Atmosphere containing flammable gases or vapors or combustible dust		
	Class I	Class II	Class III
	A B C D	E F G	none

DIVISION I EXAMPLES

Class I:
- Spray booth interiors
- Areas adjacent to spraying or painting operations using volatile flammable solvents
- Open tanks or vats of volatile flammable liquids
- Drying or evaporation rooms for flammable vents
- Areas where fats and oil extraction equipment using flammable solvents are operated
- Cleaning and dyeing plant rooms that use flammable liquids that do not contain adequate ventilation
- Refrigeration or freezer interiors that store flammable materials
- All other locations where sufficient ignitable quantities of flammable gases or vapors are likely to occur during routine operations

Class II:
- Grain and grain products
- Pulverized sugar and cocoa
- Dried egg and milk powders
- Pulverized spices
- Starch and pastes
- Potato and wood flour
- Oil meal from beans and seeds
- Dried hay
- Any other organic material that may produce combustible dusts during their use or handling

Class III:
- Portions of rayon, cotton, or other textile mills
- Manufacturing and processing plants for combustible fibers, cotton gins, and cotton seed mills
- Flax processing plants
- Clothing manufacturing plants
- Woodworking plants
- Other establishments involving similar hazardous processes or conditions

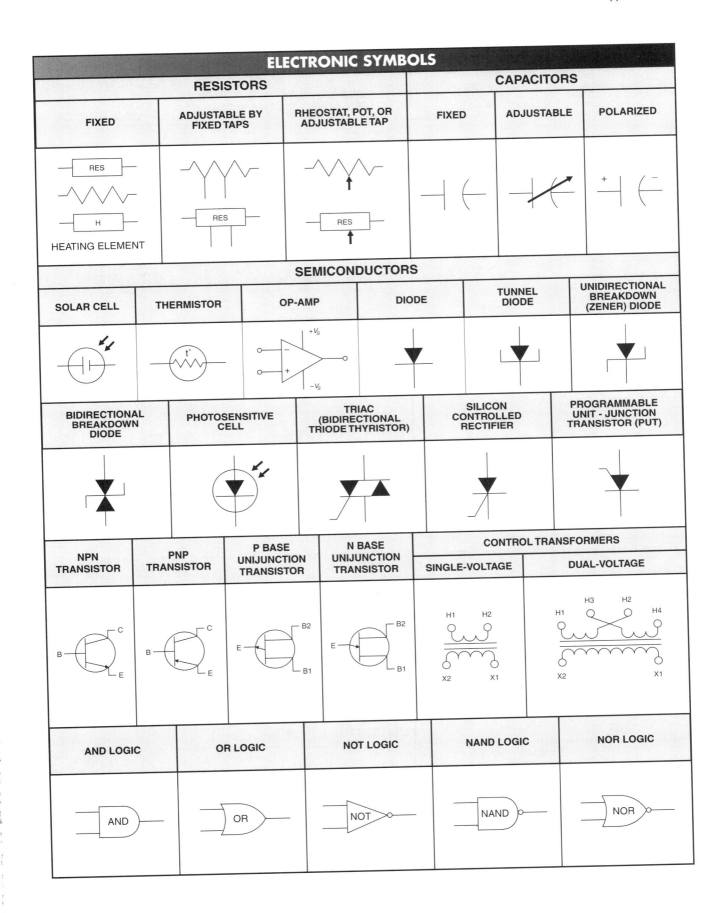

INDUSTRIAL ELECTRICAL SYMBOLS...

DISCONNECT	CIRCUIT INTERRUPTER	CIRCUIT BREAKER WITH THERMAL OL	CIRCUIT BREAKER WITH MAGNETIC OL	CIRCUIT BREAKER W/ THERMAL AND MAGNETIC OL

LIMIT SWITCHES

NORMALLY OPEN	NORMALLY CLOSED	FOOT SWITCHES	PRESSURE AND VACUUM SWITCHES	LIQUID LEVEL SWITCH	TEMPERATURE-ACTUATED SWITCH	FLOW SWITCH (AIR, WATER, ETC.)
NO	NO	NO	NO	NO	NO	NO
HELD CLOSED	HELD OPEN	NC	NC	NC	NC	NC

SPEED (PLUGGING) | ANTI-PLUG | SYMBOLS FOR STATIC SWITCHING CONTROL DEVICES

STATIC SWITCHING CONTROL IS A METHOD OF SWITCHING ELECTRICAL CIRCUITS WITHOUT USE OF CONTACTS, PRIMARILY BY SOLID-STATE DEVICES. USE SYMBOLS SHOWN IN TABLE AND ENCLOSE THEM IN A DIAMOND.

INPUT COIL | OUTPUT NO | LIMIT SWITCH NO | LIMIT SWITCH NC

SELECTOR

TWO-POSITION

	J	K
A1	X	
A2		X

X-CONTACT CLOSED

THREE-POSITION

	J	K	L
A1	X		
A2			X

X-CONTACT CLOSED

TWO-POSITION SELECTOR PUSHBUTTON

CONTACTS	SELECTOR POSITION			
	A		B	
	BUTTON		BUTTON	
	FREE	DEPRESSED	FREE	DEPRESSED
1-2	X			
3-4		X	X	X

X - CONTACT CLOSED

PUSHBUTTONS

MOMENTARY CONTACT				MAINTAINED CONTACT		ILLUMINATED
SINGLE CIRCUIT	DOUBLE CIRCUIT NO AND NC	MUSHROOM HEAD	WOBBLE STICK	TWO SINGLE CIRCUIT	ONE DOUBLE CIRCUIT	
NO / NC						

...INDUSTRIAL ELECTRICAL SYMBOLS...

CONTACTS

INSTANT OPERATING				TIMED CONTACTS - CONTACT ACTION RETARDED AFTER COIL IS:			
WITH BLOWOUT		WITHOUT BLOWOUT		ENERGIZED		DE-ENERGIZED	
NO	NC	NO	NC	NOTC	NCTO	NOTO	NCTC

OVERLOAD RELAYS

THERMAL	MAGNETIC

SUPPLEMENTARY CONTACT SYMBOLS

SPST NO		SPST NC		SPDT		TERMS
SINGLE BREAK	DOUBLE BREAK	SINGLE BREAK	DOUBLE BREAK	SINGLE BREAK	DOUBLE BREAK	SPST SINGLE-POLE, SINGLE-THROW

DPST, 2NO		DPST, 2NC		DPDT	
SINGLE BREAK	DOUBLE BREAK	SINGLE BREAK	DOUBLE BREAK	SINGLE BREAK	DOUBLE BREAK

TERMS:
- SPST — SINGLE-POLE, SINGLE-THROW
- SPDT — SINGLE-POLE, DOUBLE-THROW
- DPST — DOUBLE-POLE, SINGLE-THROW
- DPDT — DOUBLE-POLE, DOUBLE-THROW
- NO — NORMALLY OPEN
- NC — NORMALLY CLOSED

METER (INSTRUMENT)

INDICATE TYPE BY LETTER

TO INDICATE FUNCTION OF METER OR INSTRUMENT, PLACE SPECIFIED LETTER OR LETTERS WITHIN SYMBOL.

AM or A	AMMETER	VA	VOLTMETER
AH	AMPERE HOUR	VAR	VARMETER
µA	MICROAMMETER	VARH	VARHOUR METER
mA	MILLAMMETER	W	WATTMETER
PF	POWER FACTOR	WH	WATTHOUR METER
V	VOLTMETER		

PILOT LIGHTS

INDICATE COLOR BY LETTER

NON PUSH-TO-TEST	PUSH-TO-TEST

INDUCTORS

IRON CORE

AIR CORE

COILS

DUAL-VOLTAGE MAGNET COILS

HIGH-VOLTAGE	LOW-VOLTAGE
LINK — 1 2 3 4	LINKS — 1 2 3 4

BLOWOUT COIL

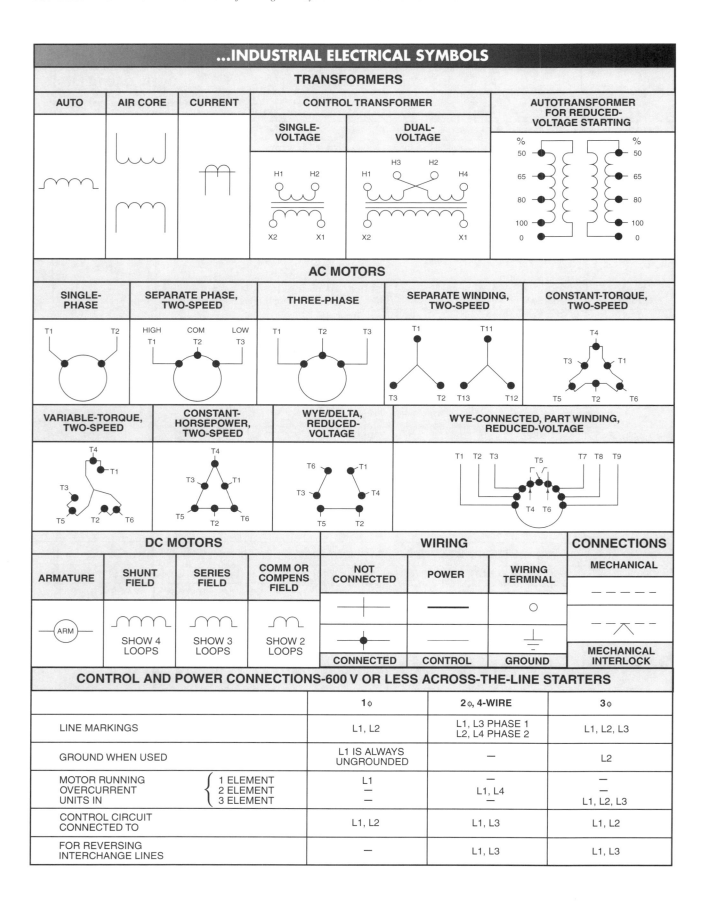

ELECTRICAL/ELECTRONIC ABBREVIATIONS/ACRONYMS

Abbr/Acronym	Meaning	Abbr/Acronym	Meaning	Abbr/Acronym	Meaning
A	Ammeter; Ampere; Anode; Armature	FU	Fuse	PNP	Positive-Negative-Positive
AC	Alternating Current	FWD	Forward	POS	Positive
AC/DC	Alternating Current; Direct Current	G	Gate; Giga; Green; Conductance	POT.	Potentiometer
A/D	Analog to Digital	GEN	Generator	P-P	Peak-to-Peak
AF	Audio Frequency	GRD	Ground	PRI	Primary Switch
AFC	Automatic Frequency Control	GY	Gray	PS	Pressure Switch
Ag	Silver	H	Henry; High Side of Transformer; Magnetic Flux	PSI	Pounds Per Square Inch
ALM	Alarm			PUT	Pull-Up Torque
AM	Ammeter; Amplitude Modulation	HF	High Frequency	Q	Transistor
AM/FM	Amplitude Modulation; Frequency Modulation	HP	Horsepower	R	Radius; Red; Resistance; Reverse
		Hz	Hertz	RAM	Random-Access Memory
ARM.	Armature	I	Current	RC	Resistance-Capacitance
Au	Gold	IC	Integrated Circuit	RCL	Resistance-Inductance-Capacitance
AU	Automatic	INT	Intermediate; Interrupt	REC	Rectifier
AVC	Automatic Volume Control	INTLK	Interlock	RES	Resistor
AWG	American Wire Gauge	IOL	Instantaneous Overload	REV	Reverse
BAT.	Battery (electric)	IR	Infrared	RF	Radio Frequency
BCD	Binary Coded Decimal	ITB	Inverse Time Breaker	RH	Rheostat
BJT	Bipolar Junction Transistor	ITCB	Instantaneous Trip Circuit Breaker	rms	Root Mean Square
BK	Black	JB	Junction Box	ROM	Read-Only Memory
BL	Blue	JFET	Junction Field-Effect Transistor	rpm	Revolutions Per Minute
BR	Brake Relay; Brown	K	Kilo; Cathode	RPS	Revolutions Per Second
C	Celsius; Capacitance; Capacitor	L	Line; Load; Coil; Inductance	S	Series; Slow; South; Switch
CAP.	Capacitor	LB-FT	Pounds Per Foot	SCR	Silicon Controlled Rectifier
CB	Circuit Breaker; Citizen's Band	LB-IN.	Pounds Per Inch	SEC	Secondary
CC	Common-Collector Configuration	LC	Inductance-Capacitance	SF	Service Factor
CCW	Counterclockwise	LCD	Liquid Crystal Display	1 PH; 1φ	Single-Phase
CE	Common-Emitter Configuration	LCR	Inductance-Capacitance-Resistance	SOC	Socket
CEMF	Counter Electromotive Force	LED	Light Emitting Diode	SOL	Solenoid
CKT	Circuit	LRC	Locked Rotor Current	SP	Single-Pole
CONT	Continuous; Control	LS	Limit Switch	SPDT	Single-Pole, Double-Throw
CPS	Cycles Per Second	LT	Lamp	SPST	Single-Pole, Single-Throw
CPU	Central Processing Unit	M	Motor; Motor Starter; Motor Starter Contacts	SS	Selector Switch
CR	Control Relay			SSW	Safety Switch
CRM	Control Relay Master	MAX.	Maximum	SW	Switch
CT	Current Transformer	MB	Magnetic Brake	T	Tera; Terminal; Torque; Transformer
CW	Clockwise	MCS	Motor Circuit Switch	TB	Terminal Board
D	Diameter; Diode; Down	MEM	Memory	3 PH; 3φ	Three-Phase
D/A	Digital to Analog	MED	Medium	TD	Time Delay
DB	Dynamic Braking Contactor; Relay	MIN	Minimum	TDF	Time Delay Fuse
DC	Direct Current	MN	Manual	TEMP	Temperature
DIO	Diode	MOS	Metal-Oxide Semiconductor	THS	Thermostat Switch
DISC.	Disconnect Switch	MOSFET	Metal-Oxide Semiconductor Field-Effect Transistor	TR	Time Delay Relay
DMM	Digital Multimeter			TTL	Transistor-Transistor Logic
DP	Double-Pole	MTR	Motor	U	Up
DPDT	Double-Pole, Double-Throw	N; NEG	North; Negative	UCL	Unclamp
DPST	Double-Pole, Single-Throw	NC	Normally Closed	UHF	Ultrahigh Frequency
DS	Drum Switch	NEUT	Neutral	UJT	Unijunction Transistor
DT	Double-Throw	NO	Normally Open	UV	Ultraviolet; Undervoltage
DVM	Digital Voltmeter	NPN	Negative-Positive-Negative	V	Violet; Volt
EMF	Electromotive Force	NTDF	Nontime-Delay Fuse	VA	Volt Amp
F	Fahrenheit; Fast; Field; Forward; Fuse	O	Orange	VAC	Volts Alternating Current
FET	Field-Effect Transistor	OCPD	Overcurrent Protection Device	VDC	Volts Direct Current
FF	Flip-Flop	OHM	Ohmmeter	VHF	Very High Frequency
FLC	Full-Load Current	OL	Overload Relay	VLF	Very Low Frequency
FLS	Flow Switch	OZ/IN.	Ounces Per Inch	VOM	Volt-Ohm-Milliammeter
FLT	Full-Load Torque	P	Peak; Positive; Power; Power Consumed	W	Watt; White
FM	Fequency Modulation	PB	Pushbutton	w/	With
FREQ	Frequency	PCB	Printed Circuit Board	X	Low Side of Transformer
FS	Float Switch	PH;	Phase	Y	Yellow
FTS	Foot Switch	PLS	Plugging Switch	Z	Impedance

NONLOCKING WIRING DEVICES

2-POLE, 3-WIRE

Wiring Diagram	NEMA ANSI	Receptacle Configuration	Rating
	5-15 C73.11		15 A 125 V
	5-20 C73.12		20 A 125 V
	5-30 C73.45		30 A 125 V
	5-50 C73.46		50 A 125 V
	6-15 C73.20		15 A 250 V
	6-20 C73.51		20 A 250 V
	6-30 C73.52		30 A 250 V
	6-50 C73.53		50 A 250 V
	7-15 C73.28		15 A 277 V
	7-20 C73.63		20 A 277 V
	7-30 C73.64		30 A 277 V
	7-50 C73.65		50 A 277 V

3-POLE, 3-WIRE

Wiring Diagram	NEMA ANSI	Receptacle Configuration	Rating
	10-20 C73.23		20 A 125/250 V
	10-30 C73.24		30 A 125/250 V
	10-50 C73.25		50 A 125/250 V
	11-15 C73.54		15 A 3φ 250 V
	11-20 C73.55		20 A 3φ 250 V
	11-30 C73.56		30 A 3φ 250 V
	11-50 C73.57		50 A 3φ 250 V

3-POLE, 4-WIRE

Wiring Diagram	NEMA ANSI	Receptacle Configuration	Rating
	14-15 C73.49		15 A 125/250 V
	14-20 C73.50		20 A 125/250 V
	14-30 C73.16		30 A 125/250 V
	14-50 C73.17		50 A 125/250 V
	14-60 C73.18		60 A 125/250 V
	15-15 C73.58		15 A 3φ 250 V
	15-20 C73.59		20 A 3φ 250 V
	15-30 C73.60		30 A 3φ 250 V
	15-50 C73.61		50 A 3φ 250 V
	15-60 C73.62		60 A 3φ 250 V

4-POLE, 4-WIRE

Wiring Diagram	NEMA ANSI	Receptacle Configuration	Rating
	18-15 C73.15		15 A 3φY 120/208 V
	18-20 C73.26		20 A 3φY 120/208 V
	18-30 C73.47		30 A 3φY 120/208 V
	18-50 C73.48		50 A 3φY 120/208 V
	18-60 C73.27		60 A 3φY 120/208 V

Appendix 303

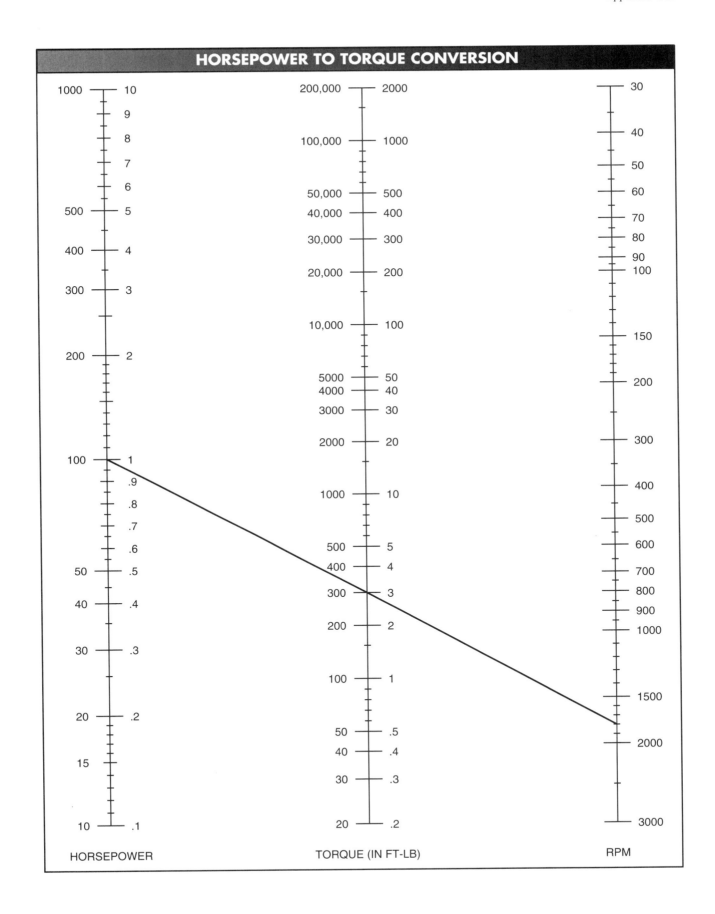

COMMON SERVICE FACTORS

Equipment	Service factor
Blowers	
Centrifugal	1.00
Vane	1.25
Compressors	
Centrifugal	1.25
Vane	1.50
Conveyors	
Uniformly loaded or fed	1.50
Heavy-duty	2.00
Elevators	
Bucket	2.00
Freight	2.25
Extruders	
Plastic	2.00
Metal	2.50
Fans	
Light-duty	1.00
Centrifugal	1.50
Machine tools	
Bending roll	2.00
Punch press	2.25
Tapping machine	3.00
Mixers	
Concrete	2.00
Drum	2.25
Paper mills	
De-barking machines	3.00
Beater and pulper	2.00
Bleacher	1.00
Dryers	2.00
Log haul	2.00
Printing presses	1.50
Pumps	
Centrifugal—general	1.00
Centrifugal—sewage	2.00
Reciprocating	2.00
Rotary	1.50
Textile	
Batchers	1.50
Dryers	1.50
Looms	1.75
Spinners	1.50
Woodworking machines	1.00

Certificate of Completion

in
Motor Control Principles

has received instruction and completed activities in the following areas

___

- ____ 1-1: Electrical Prefixes
- ____ 1-2: Using Ohm's Law and the Power Formula
- ____ 1-3: Resistance in Series Circuits
- ____ 1-4: Resistance in Parallel Circuits
- ____ 1-5: Resistance in Series/Parallel Circuits
- ____ 1-6: Resistor Color Coding
- ____ 1-7: Finding Total Resistance of Resistors Connected in Parallel Using a Calculator
- ____ 2-1: Multimeter Use
- ____ 2-2: Temperature Conversion
- ____ 2-3: Reading Graphs
- ____ 2-4: Applying Electrical Principles when Troubleshooting
- ____ 3-1: Electrical Glove Selection
- ____ 3-2: Grounding
- ____ 3-3: Safety Label Information
- ____ 3-4: Troubleshooting Ground Problems
- ____ 4-1: Symbol and Abbreviation Identification
- ____ 4-2: Wiring and Line Diagrams
- ____ 4-3: Allowable Voltage Drop across Conductors
- ____ 4-4: Troubleshooting Power and Control Circuits
- ____ 5-1: Numerical Cross-Reference Numbers
- ____ 5-2: Assigning Wire-Reference (Terminal) Numbers
- ____ 5-3: Manufacturer Terminal Numbers
- ____ 5-4: Wiring Control Panels
- ____ 5-5: Basic Switching Logic
- ____ 5-6: Troubleshooting Control Circuits

_____
Program Instructor

_____
Institution

_____
Date

Certificate of Completion

in
Motor Control Devices

has received instruction and completed activities in the following areas

---

___ 6-1: Solenoids
___ 6-2: Directional Control Valves
___ 6-3: Fluid Power Color Coding
___ 6-4: Troubleshooting DC Motor Circuits

___ 7-1: AC Generators
___ 7-2: Transformers
___ 7-3: Troubleshooting AC Motors and Motor Circuits

___ 8-1: Ambient Temperature Compensation with Overloads
___ 8-2: Overload Trip Time
___ 8-3: Overload Heater Size
___ 8-4: Contactor and Motor Starter Ratings
___ 8-5: Sizing Motor Protection, Motor Starters, and Wire
___ 8-6: Motor Starter Replacement Parts
___ 8-7: Motor Drives
___ 8-8: Troubleshooting Contactors

___ 9-1: Enclosure Selection
___ 9-2: Alternating Motor Control
___ 9-3: Level Control
___ 9-4: Temperature Control
___ 9-5: Selecting Blowers and Exhaust Fans
___ 9-6: Troubleshooting Control Device Circuits

_____ _____ _____
Program Instructor   Institution   Date

Certificate of Completion

in

Motor Control Circuits

has received instruction and completed activities in the following areas

- 10-1: Reversing Motor Circuits
- 10-2: Reversing Three-Phase Motors
- 10-3: Reversing Single-Phase Motors
- 10-4: Reversing Dual-Voltage Motors
- 10-5: Reversing DC Motors
- 10-6: Troubleshooting Reversing Motor Circuits
- 10-7: Hard Wiring Reversing Circuits
- 10-8: Reversing Circuits and Terminal Strips
- 10-9: Reversing Circuits and PLCs
- 11-1: Wye and Delta Transformer Configurations
- 11-2: Motor Control Centers
- 11-3: Busway Systems
- 11-4: Troubleshooting Power Circuits
- 11-5: 120/240 V, Single-Phase Systems
- 11-6: 120/208 V, Three-Phase Systems
- 11-7: 120/240 V, Three-Phase Systems
- 11-8: 277/480 V, Three-Phase Systems
- 11-9: Plug and Receptacle Configurations and Ratings
- 12-1: Electronic Device Symbols
- 12-2: Digital Circuit Logic Functions
- 12-3: Combination Logic Circuits
- 12-4: Solid-State Relays and Switches
- 12-5: Photovoltaic Cells
- 12-6: Troubleshooting Digital Circuits
- 13-1: ON-Delay Timer Applications
- 13-2: OFF-Delay Timer Applications
- 13-3: One-Shot Timer Applications
- 13-4: Recycle Timer Applications
- 13-5: Combination Timing Logic Applications
- 13-6: Selecting and Setting Timers
- 13-7: Troubleshooting Timer Circuits
- 14-1: Relays
- 14-2: Heat Sink Selection
- 14-3: Heat Sink Installation
- 14-4: Solid-State Relay Installation
- 14-5: Troubleshooting at Motor Starters
- 15-1: Proximity Sensors
- 15-2: Proximity Sensor Installation
- 15-3: Determining Activating Frequency
- 15-4: Applying Photoelectric Sensors
- 15-5: Troubleshooting Photoelectric Sensors

_____ _____ _____
Program Instructor   Institution   Date

Certificate of Completion

in
Motor Control Methods

_____

has received instruction and completed activities in the following areas

_____

- 16-1: Programmable Controller Input and Output Identification
- 16-2: Programmable Controller Input and Output Connections
- 16-3: Alarm Output Connection
- 16-4: Developing PLC Programmable Circuits
- 16-5: Troubleshooting PLC Inputs and Outputs

- 17-1: Primary Resistor Reduced-Voltage Starting
- 17-2: Part-Winding Reduced-Voltage Starting
- 17-3: Autotransformer Reduced-Voltage Starting
- 17-4: Wye/Delta Reduced-Voltage Starting
- 17-5: Closed Transition Reduced-Voltage Starting
- 17-6: Troubleshooting Reduced-Voltage Starting Circuits

- 18-1: One-Direction Motor Plugging
- 18-2: Two-Direction Motor Plugging
- 18-3: Two-Speed Separate Winding Motors
- 18-4: Two-Speed Consequent Pole Motors
- 18-5: Motor Torque and Horsepower
- 18-6: Troubleshooting Two-Speed Circuits
- 18-7: Troubleshooting Two-Direction Plugging Circuits
- 18-8: Troubleshooting Electronic Braking Circuits

- 19-1: Conveyor Drive Methods
- 19-2: Motor Coupling Selection
- 19-3: Extension Cord Selection
- 19-4: Load Variations

_____ _____
Program Instructor   Institution

_____
Date